POWDER METALLURGY

Applications, Advantages and Limitations

POWDER METALLURGY

Applications, Advantages and Limitations

Edited by

ERHARD KLAR
Manager, Particle Technology
Glidden Metals Chem./Met. Division
SCM Corporation

A project of the Powder Metallurgy Committee of the Mechanical Working and Forming Division of American Society for Metals

AMERICAN SOCIETY FOR METALS
Metals Park, Ohio 44073

Library of Congress Catalog Card No. 82-73609

ISBN: 0-87170-154-5

SAN 204-7586

Editorial and production coordination by
Carnes Publication Services, Inc.

PRINTED IN THE UNITED STATES OF AMERICA

Preface

In spite of the impressive growth of the powder metallurgy industry since World War II, it has become increasingly clear that many metallurgists and design engineers at end user companies have little knowledge of the P/M process and its capabilities. Recognizing this deficiency, the Powder Metallurgy Committee of the Mechanical Working and Forming Division of ASM has over the past three years presented a one-day seminar on powder metallurgy to several ASM Chapters across the nation. This book is the outcome of that highly successful seminar.

In its expanded version the book is, like the seminar, neither elementary nor advanced. Although we cannot make any claim to completeness, we believe that the book will familiarize the reader with an important cross section of the powder metallurgy industry, with particular emphasis on its applications and its advantages and limitations. Thus, in addition to an introductory overview and an exposure to the basics of compaction, sintering and P/M designing, the selected topics include powder metallurgy of ferrous alloys (for the automotive industry), stainless steels, copper alloys, magnetic materials, aluminum alloys and aerospace materials.

The authors of the papers presented at the seminar were, in most instances, lecturers with many years of experience in their respective fields. This book provides richly illustrated, updated and more detailed accounts of the various topics, and it is believed that it will be of value also to the practicing powder metallurgist.

ERHARD KLAR

Contents

1 **Introduction** *Donald Gustafson* 1

2 **Elements of Powder Metallurgy** *Darrell W. Smith* 7

Manufacture of Metal Powders 7
Characteristics of Metal Powders 9
Particle Size and Size Distribution 11
Particle Shape 14
Apparent Density 17
Flow Rate 18
Compressibility 19
Compaction of Powders 21
Compaction Techniques 22
Powder Behavior During Compaction 25
Strength of Powder Compacts 28
Sintering 29
Development of Mechanical Properties 30
Development of Microstructure 32
Sintering Mechanisms 36
Alloying by P/M Techniques 38

3 **Designing for Powder Metals** *Ron Sadlon* 43

Pressing Equipment 43
Compaction of Powders 45
Tooling 45
Design of Powder Metal Parts 47
Part Size and Shape 47
Length-to-Height Ratio 48
Splines 49
Knurls and Serrations 49

Chamfers and Bevels 49
Corner Reliefs 50
Shapes That Cannot Be Molded 50
Tolerances 50
Surface Finish 51
Additional Processing of Powder Metal Parts 51

4 Ferrous Powder Metallurgy **55**
Part I: Pressed and Sintered Parts and Their Applications
Peter W. Lee **55**
Ferrous Metal Powders 56
Direct Reduction of Oxides 56
Atomization of Liquid Metal 60
Compaction and Sintering Technology 62
The Compaction Process 62
The Sintering Process 66
Typical Properties of P/S Parts 69
Typical Applications of P/S Parts 73
High-Density Ferrous Powder Metallurgy 83
Part II: Fully Dense Parts and Their Applications
B. Lynn Ferguson **86**
P/M Forging Technology 86
Preform Production 86
Preform Design 87
The Forging Step 95
Effects of Forging Variables on Mechanical Properties 97
Applications 99
P/M Specialty Steels 112
P/M Tool Steels 113
Processing of P/M Tool Steels 118
Ferrous Superalloys 119

5 Stainless Steel Powder Metallurgy
Donald L. Dyke and Harry D. Ambs **123**
Powders 123
Sintering 129
Corrosion Resistance 135
Applications 138
Structural Parts 138
Porous Parts 141
Future Prospects 142

6 P/M Copper and Copper Alloys
Erhard Klar and Paul E. Matthews **145**

Powder Production 145
Atomization 146
Electrolysis 149
Oxide Reduction 150
Hydrometallurgy 150
Applications 151
Self-Lubricating Porous Bronze Bearings 151
Friction Materials 158
P/M Copper-Base Structural Parts 163
Electrical Parts 169
Copper-Infiltrated Parts 172
Other Applications 182

7 The Effects of Phosphorus on the Properties of Iron Alloys for Magnetic Applications *Kenneth H. Moyer* **185**

The Effects of Phosphorus on DC Properties 186
Full-Density Parts 188
Porous Compacts 193
Aging 199

8 Aluminum P/M — Properties and Applications
John D. Generous and Wayne C. Montgomery **211**

Material Characteristics 211
Properties and Performance 212
Corrosion Resistance 214
Light Weight 215
Conductivity 215
Manufacturing Process 216
Compacting 216
Sintering 218
Secondary Operations 219
Sizing, Coining and Cold Forming 219
Heat Treatment 220
Machining 220
Finishing 221
Joining 222
Applications 224
Business Machines 224
Automotive Components 227

Appliances 229
Power Tools 229
Miscellaneous Applications 231

9 Wrought Aluminum P/M Alloys *Robert H. Graham* **235**

Typical Tensile Properties 236
Corrosion Resistance 236
Fracture Toughness/Strength Relationships 238
Notched Fatigue Strength 238
Nominal Compositions 238
Grain Structure 239
Physical Properties 242
Manufacture of P/M Alloys 242

10 Powder Metallurgy Parts for Aerospace Applications
J. H. Moll, V. C. Petersen, and E. J. Dulis **247**

P/M Process 248
P/M Superalloys 258
Powder Characteristics 258
Mechanical Properties 268
P/M Superalloy Applications 275
P/M Titanium Alloys 279
Powder Characteristics 279
Mechanical Properties of Prealloyed Powder 283
Mechanical Properties of Blended Elemental Powder 289
P/M Titanium Applications 291

Index **299**

Chapter 1
Introduction

Donald Gustafson
Director, International
Imperial Clevite Inc.

Powder metallurgy (P/M) has the distinction of being at the same time one of the oldest and one of the most modern methods known for the fabrication of metal articles. In prehistoric times, powder metallurgy techniques were used to process metals with melting points above those attainable (or, in some instances, practical) by means of the technology which then existed. Development of these techniques probably began with iron in the form of sponge produced by reduction of iron oxide in charcoal furnaces. This material was subsequently forged into solid iron or steel. The ancients produced some rather extraordinary materials by utilizing these methods. The Arabs, for instance, made their remarkable swords by employing such a technique.

Powder materials were first press-bonded in the early 1800's in a manner similar to that employed today. It was in the 1920's that the process was really first used commercially with the development of the porous bronze bushing and the related techniques for mass production.

While this volume will present powder metallurgy in its most conventional form—that is, as a method of producing discrete shapes from metal powders—powder metallurgy in its broader sense covers a much wider spectrum of modern technology. For example (and this is only a partial list):

- The "gold" or "silver" printing used on labels, on greeting cards and on a variety of other printed matter is achieved through the use of finely divided, specially prepared bronze and aluminum powders.
- The cemented carbides almost universally used in cutting tools and in many high-wear applications are prepared from carbide and cobalt or from similar finely divided materials.
- Many fuel cell and other energy generation or storage devices, including the nickel-cadmium rechargeable battery, depend on metal powders as the basis for manufacture of the porous components necessary for their function.
- Floating soap and lightweight concrete are both dependent for these properties on the addition of aluminum powder which reacts chemically with the base material to form tiny bubbles of hydrogen which are trapped in the structure.
- Catalysts, vital in petroleum refining as well as in many other chemical production processes, are generally made from metal powders.
- Flour for baking, as well as foods such as breakfast cereals, are iron enriched by direct addition of specially prepared iron powder.
- The familiar incandescent lamp has utilized tungsten filaments made from powder almost since its inception.

Powder metallurgy, as it is practiced today, has expanded to become a very pervasive technology. While not large in the sense of tons of material processed when compared with casting or forging, for instance, P/M is nonetheless an extremely important industrial technique. It is very safe to paraphrase a radio commercial of some years back and say, "You're never very far from a product of powder metallurgy". An automobile, for instance, may employ upwards of 50 P/M parts. P/M components can be found in washing machines, refrigerator and air conditioning compressors, bicycles and lawnmowers. Farm machinery and industrial hydraulic equipment are large users. Data-processing equipment, office copiers, postage meters and similar machines may actually have more than a hundred P/M parts designed into them. For the outdoorsman, the fishing reel and the firearm both are possible applications of powder metallurgy. The watch you wear may have a P/M case. Tape decks and phonograph turntables, as well as TV sets, also utilize components made by powder metallurgy.

In addition, aircraft engines depend on P/M for several of the high-performance alloys which allow them to operate safely and more efficiently. Tool steels have also seen significant improvements made possible through P/M. We could continue to extend this catalog of usage, but the point has been made: P/M has become an important technology to a large segment of industry.

In the following chapters we will present just enough theory to provide an appreciation of some of the basics of the process. We will then move to a discussion of some of the practices involved in the fabrication of parts from metal powders. The things which can and cannot be accomplished via this process will be discussed, and the reader will be familiarized with some of the design considerations involved in the evolution of a successful application. Subsequently, a number of different materials and industrial applications will be presented as well as an insight into the world of superalloys and the methods utilized in their consolidation. Nearly everything will be directed toward that segment of powder metallurgy which concerns itself with the production of discrete shapes from metal powders.

The P/M manufacturing process can be looked at as consisting of three basic operations:

- Powder preparation
- Powder consolidation
- Sintering.

The production of raw materials—metal powders and additives—is a subject which in itself would require more space than we have available. It will be touched on briefly a time or two but not in any depth. Normally, the parts producer purchases all his raw materials, and thus his involvement begins with preparation of the materials for processing.

The comments and descriptions which follow are basically applicable to ferrous powder metallurgy, which is the largest and economically most important sector of the conventional P/M business. Almost all of this information, however, is also applicable to other materials, because the process is very similar for most metals and their alloys.

There are several things which the parts producer accomplishes in the powder preparation phase of his operation. The most obvious (assuming, of course, that he is going to make iron or steel parts) is the combining of the iron powder with the other necessary ingredients. All materials are added as weight percentages, and they can include graphite, and in many instances copper and/or nickel. A very typical P/M mix is iron plus 2% copper and 1% graphite. An internal lubricant, such as zinc stearate or a stearate derivative, will also be used in amounts from 1/2 to 1% or more, depending on the application. The purpose of such a lubricant is to facilitate ejection of the compact from the tooling and to prevent scoring of the punches, dies and core rods.

The powders themselves are highly engineered materials, and great care must be exercised in preparing them for the compacting operation. Additions, of course, must be precisely weighed. Mixing, usually done in double cone blenders in lot sizes ranging from a few hundred to tens of thousands of pounds, is carefully controlled. Customarily, several checks are made on powder samples taken from the blender before the material is removed. The important properties analyzed include apparent density, flow rate, compressibility, sinter size change, and, possibly, sintered strength and hardness. If problems are found, corrections can be made before the mix is removed from the blender.

The mixed powder is then moved to the production floor. The compacting or briquetting operation is accomplished in specially designed mechanical or hydraulic presses. The presses are fitted with very precise, closely toleranced tooling which has been specifically built for the part to be made. The mixed powder is brought to a position directly above the die by a feed mechanism, and the proper amount of material, determined by the volume of the die cavity, is allowed to flow into the tooling. The feed mechanism is then retracted, leveling the powder. Pressure is applied to the powder mass by top and bottom punches simultaneously. The force applied to ferrous powders will be in the range of 35 tons per square inch of the plan area of the part. As an example, a simple ring with a 3-in. O.D. and a 1/2-in. wall has a plan area of just under 4 in.^{2}. This ring, if pressed to a standard density, would require a total force of about 140 tons.

The last of the three basic operations is sintering. In this step the compact, or "green" part, is subjected to high temperatures in the presence of a protective atmosphere. Standard sintering temperatures for iron and steel parts are in the neighborhood of 1120°C, although this may vary from a few degrees lower to 100 to 150° higher. The ovens, or furnaces, utilized may be of any of several types that are available. Continuous mesh belt furnaces can be utilized to about 1150°C, although pusher and walking beam units can be used in this temperature range and are necessary for higher temperatures. Box, bell, vacuum and other types of batch ovens are also utilized on occasion, but the continuous mesh belt unit is probably the most prevalent. The protective atmospheres conventionally used are endothermic gas, generated from natural gas or propane, and dissociated ammonia. As energy costs rise, atmospheres based on bulk nitrogen are becoming more popular.

The sintering operation, through a combination of processes which occur within the compact as it is subjected to high temperatures, is the step that provides the final strength to the material. The relative properties of the "green" and "sintered" materials will be discussed and explained in subsequent chapters.

These manufacturing steps—blending, compacting and sintering—are the basics of the production of P/M parts. Any number of subsequent operations may be performed on the sintered compact, depending on the requirements of the finished part.

Powder metallurgy is very much a "net shape" or "near net shape" process, depending on the actual design features and tolerances required in the finished part. Because a

P/M part is restricted to a geometry which allows ejection from the tooling used to shape it, certain features such as undercuts and cross holes must be provided in secondary operations. Frequently, too, tolerances more restrictive than those attainable in the basic process must be provided for through standard machining operations following sintering. On that subject, P/M can regularly provide dimensional control which allows tolerances of ±0.1 mm in the direction of pressing and ±0.05 mm or better, depending on over-all size, in the radial direction.

To the designer who is not completely familiar with powder metallurgy, the P/M process may seem difficult to work with and to specify. In reality, it need be no different in these respects from more familiar fabrication methods. As a matter of fact, it can simplify production in many instances, since P/M allows many design freedoms which are not available from other metal forming techniques. For the uninitiated, such concepts as "variable density", "copper infiltration", "apparent hardness" and several others which are somehow peculiar to the process can be intimidating. Sound advice to the designer is to ignore all these and to let the P/M parts fabricator concern himself with these details. What the partsmaker needs from his customer is the same design information necessary for any other metalworking technique—required material properties (strength, hardness, etc.), application details, special environmental considerations, assembly information and any known special requirements. Reputable P/M parts suppliers are well equipped to translate such requirements into the jargon peculiar to the technology and to discuss this with the user in the depth of detail required.

The material that follows will explain several of the concepts peculiar to P/M and then will cover in some detail the practical aspects of designing for and using the products of the process.

Chapter 2

Elements of Powder Metallurgy

Darrell W. Smith
Department of Metallurgical Engineering
Michigan Technological University

This chapter is the essence of a lecture whose purpose is to introduce the rudimentary technical concepts of powder metallurgy (P/M) to an audience generally familiar with materials processing technologies, but not with P/M specifically. As such, the text is very general in nature, but does serve to introduce P/M concepts and terminology in preparation for some of the more specific discussions which follow in this volume.

We will begin with a brief mention of the most important processes for producing metal powders and then will move to an overview of some of the characteristics of metal powders which influence their behavior during subsequent processing by standard P/M techniques. In a very simplified way, the compaction and sintering of metal powders are examined with emphasis on conventional P/M practice, specialized techniques being mentioned only when necessary for comparison or clarity. Much of the content of the following sections is covered in considerably more detail in the well-known texts by Hirschhorn[1] and Lenel[2]. These books are recommended for a more complete treatment of P/M fundamentals than is possible here.

Manufacture of Metal Powders

Metal powders for P/M products may be produced by several techniques, each of which typically belongs to one or more of the following categories: physical methods, chemical methods and mechanical methods.

A very prominent technique called "atomization" is a method for physically producing powder by solidification of droplets from a disrupted stream of the molten metal. Either a liquid, such as water, or a gas, such as argon or helium, may be employed to form the droplets, the choice being dependent on several factors, including the type of metal being atomized and the desired shape of the resulting powder particles. Atomization can be used to produce powders from a wide range of pure metals as well as from ferrous and nonferrous alloys such as stainless steels, nickel-base superalloys and titanium alloys. A schematic illustration of one form of the atomizing process is given in Fig. 1. Reviews of metal powder atomization procedures are becoming abundant in the literature,

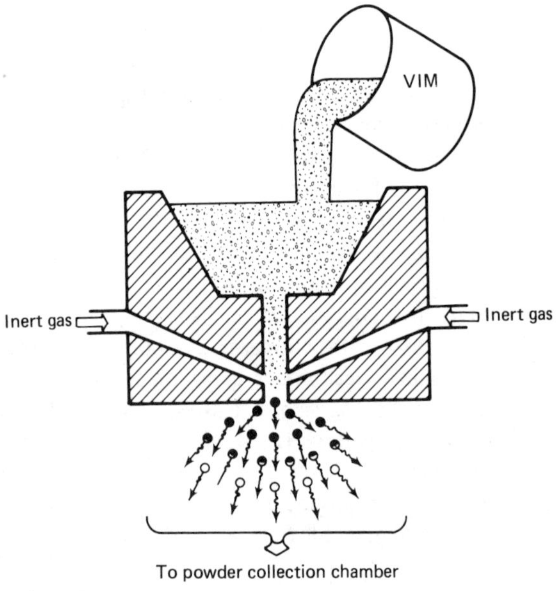

Fig. 1. Schematic illustration of inert-gas atomization of vacuum induction melted (VIM) metal. (Courtesy of J. E. Smugeresky, Sandia National Laboratories)

and the presentations by Gummeson[3] and Klar and Shafer[4] provide good introductory reading.

Many metal powders are routinely produced by chemical decomposition of a compound of the metal. Notable in this respect is the reduction of particulate metal oxides in the presence of gaseous reducing agents such as hydrogen or carbon monoxide. Powders of tungsten, molybdenum, copper and iron are manufactured by this technique. Included in the general family of reduction processes is the cathodic deposition of a metal powder (e.g., copper or silver) in an electrolytic cell containing the appropriate metal salt in the electrolyte.

Some metal compounds may be quite easily decomposed by heating to modest temperatures. Both nickel and iron powders can be produced by such "thermal decomposition" of their respective carbonyls [($Ni(CO)_4$, $Fe(CO)_5$]. Titanium metal powder can be generated by decomposition of titanium hydride.

Comminution methods (mechanical crushing and/or grinding) are essential secondary processes in the production of several metal powders. Some reduction techniques result in the formation of a spongy mass which must be crushed into particulate form. So-called sponge iron is a comminuted powder. As a primary metal powder manufacturing process, mechanical disintegration is restricted in its usefulness. Due to their inherent ductility, some metals cannot be conveniently comminuted to usable powder. Further, powders produced by comminution of brittle metals (e.g., polycrystalline tungsten) are difficult to compact at ambient temperature.

Table 1 provides a partial summary of the methods by which some of the more important powders used in P/M processing are manufactured.

Characteristics of Metal Powders

The ability of a metal powder to be effectively compacted, and the resulting properties of the compact before and after sintering, are influenced to a substantial degree by the characteristics of the starting powder. The size and shape, internal microstructure, chemical

Table 1. Processes generally used for manufacture of various metal powders

Powder material	Atomization	Chemical reduction	Comminution	Electrolytic deposition	Thermal decomposition	Precipitation from liquid or gas
Aluminum/Al alloys	X					
Beryllium			X	X		
Cobalt		X				
Copper	X	X		X		X
Copper alloys:						
Cu-Al	X					
Cu-Pb	X					
Cu-Sn (bronze)(a)	X					
Cu-Zn (brass)	X					
Cu-Ni-Zn (nickel silver)	X					
Iron	X	X	X	X	X	
Iron alloys:						
Low-alloy steel	X					
Stainless steel	X					
Tool steel	X					
Molybdenum		X				
Nickel		X			X	X
Nickel alloys	X		X			
Silver	X			X		X
Tantalum		X		X		
Tin	X					
Titanium	X	X			X	
Tungsten		X				
Zirconium					X	

(a) Bronze P/M parts are also produced from blends of elemental Cu and Sn powders.

composition and surface condition of the individual particles all play key roles in determining the static bulk properties of the powder as well as its dynamic behavior associated with flow, compaction and sintering. Because the scope of this discussion does not permit an adequate examination of these various cause-and-effect relationships, we will be content with limited consideration of the following basic powder characteristics:

- Particle size and size distribution
- Particle shape
- Apparent density

- Flowability
- Compressibility.

Particle Size and Size Distribution

Metal powders suitable for processing into P/M parts generally range in size from 0.1 to 1000 μm as the extreme limits (1 $\mu m \cong 0.00004$ in.). Normally, the size of a particle is specified by some linear dimension (e.g., "diameter"). However, most particles of real powders are not spherical, and the "particle size" then is not a quantity which can be directly measured conveniently. Usually, indirect methods are employed to determine particle size. Typical techniques involve sedimentation, turbidimetry (light attenuation), permeability (Fisher Subsieve Sizer), elutriation (Roller Analyzer), surface adsorption (BET), electrolytic resistivity (Coulter Counter) and simple sieving methods. These methods are introduced in the article by Smith[5] and are treated in detail elsewhere[6,7,8].

As suggested by the schematic particle size spectrum in Fig. 2, powders for P/M use can be categorized as either "sieve" (>44 μm, +325 mesh) or "subsieve" (<44 μm, −325 mesh). In a real powder mass, not all particles will be of the same size, and this can be quantified by determining the "particle size distribution".

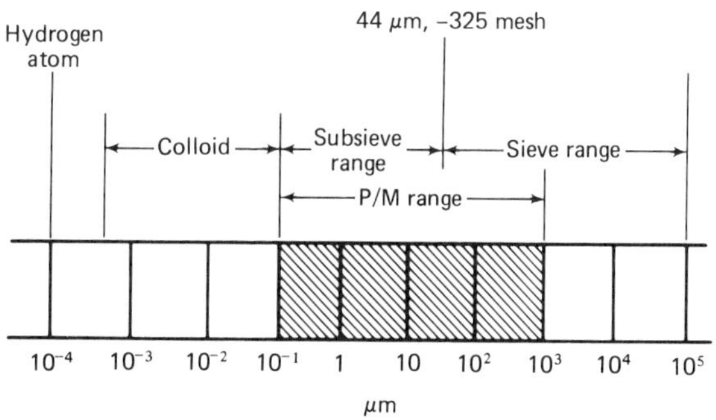

Fig. 2. Particle size spectrum.

Fig. 3. Scanning electron micrograph of molybdenum powder. Average particle size, approximately 3 μm.

Figure 3 is a scanning electron micrograph of molybdenum powder (produced by hydrogen reduction of oxides) which contains particles ranging in size from about 0.5 to 15 μm.

Particle size distributions usually are illustrated by plotting particle size versus the weight fraction of particles having that size. Depending on powder type, there are numerous possible types of distributions, some of which are illustrated in Fig. 4. A particle size measurement technique such as sieving provides data on weight fractions of particles having sizes spanning certain ranges. It is often convenient to plot such data in the form of a histogram such as that shown in Fig. 5. Other methods may provide cumulative particle size data. Figure 6 shows a simple unimodal particle size distribution and its corresponding cumulative distribution curve.

There are no clear scientific rules that dictate what type of particle size distribution is needed for a specific P/M application. However, both particle size and particle shape exert considerable influence on the behavior of a powder during subsequent processing.

In general for the P/M process, many of the properties of the powder compact and the final sintered part are

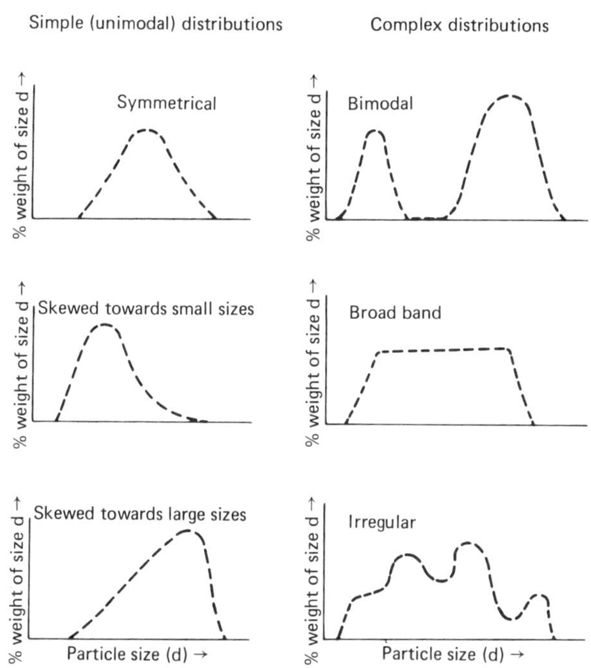

Fig. 4. Schematic illustrations of various types of particle size distributions.[1]

directly related to the extent to which powder particles establish contact with their neighbors. This, in turn, depends on both the size and shape of the particles. Some general observations relative to particle size effects include the following:

- A fine particle size is most often desirable for compaction of hard or brittle powders such as W, Al_2O_3, etc., because the increased interparticle friction helps develop sufficient adhesion for subsequent handling.
- Fine powders provide many interparticle contacts. This promotes subsequent sintering but makes it difficult to achieve uniform compacted density in large sections.
- Coarse powders result in more uniform densification during compaction but, due to fewer interparticle contacts and more sluggish sintering behavior, large pores are

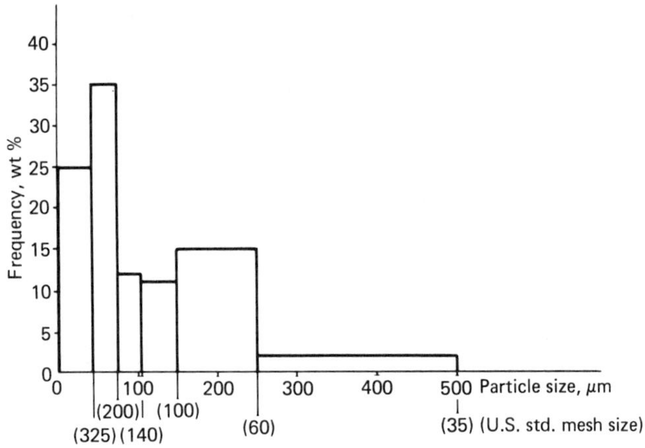

Fig. 5. Histogram representing particle size distribution of water-atomized stainless steel powder. Distribution determined by sieving. (Courtesy of E. Klar and J. Fesko, Glidden Metals)

often retained after sintering. This is detrimental to mechanical properties.

Particle Shape

The shapes of powder particles vary greatly depending on the metal and the production method employed and may play a dominant role in establishing many of the properties of the powder mass, such as packing efficiency, flowability, compressibility, etc. Particle shape greatly influences the extent of interparticle contact developed within the powder mass and, thus, the magnitude of the friction associated with the particle surfaces. The International Organization for Standardization has defined the following qualitative particle shape classifications in standard ISO 3252. These particle shapes are illustrated in the sketches in Fig. 7.

- Acicular: needle shaped
- Angular: roughly polyhedral with sharp edges
- Dendritic: branched, crystalline shape
- Fibrous: regularly or irregularly threadlike

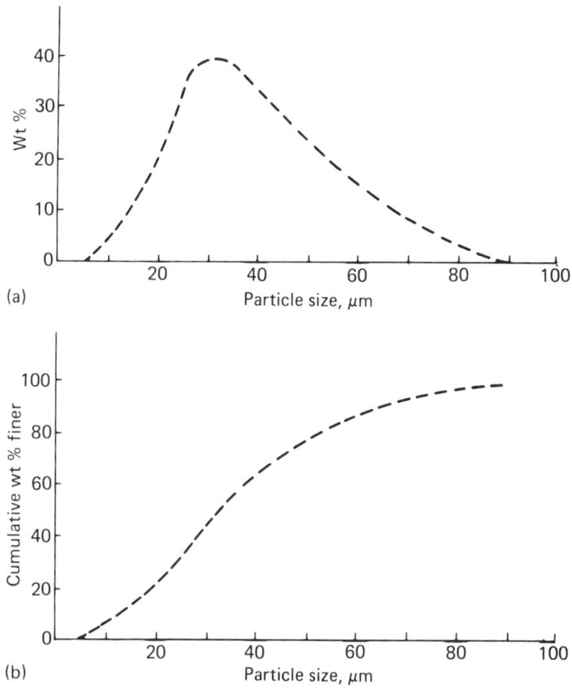

Fig. 6. Particle size distribution curves for a hypothetical powder.[9] (a) Discrete weight-size plot. (b) Cumulative plot.

- Flaky: platelike
- Granular: irregular but approximately equidimensional
- Irregular: lacking any symmetry
- Nodular: of rounded, irregular shape
- Spheroidal: nominally spherical.

A quantitative definition of actual particle shapes can be very difficult to achieve and often involves the assessment of one or more "shape factors" from microscopic study of projected particle images. Several elegant presentations of the techniques and analysis of particle shape characterization have been recently published in the book edited by Beddow and Meloy[10].

In most particle size measurement techniques, the particle shape is assumed to be generally spherical. Obviously,

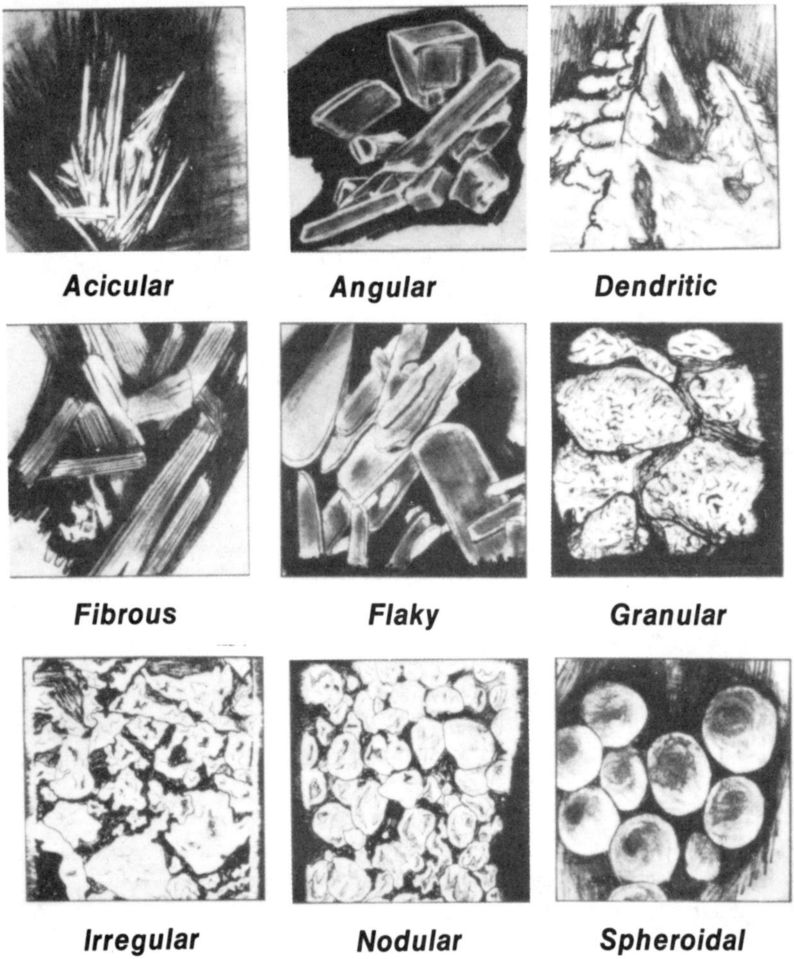

Fig. 7. Illustrations of nine categories of particle shape.[5]

spherical particles are the exception rather than the rule. Any deviation from sphericity serves to increase the specific surface area of the powder and, therefore, to increase the interparticle friction. In turn, this affects the ease with which particles can move during powder flow, when settling in a mold or die and when under pressure during compaction. Increased particle surface area (smaller particle size, more irregular particle shape, rougher particle surfaces) also promotes greater chemical reactivity of the powder. This leads to greater adsorption of atmospheric gases and water vapor, resulting in the formation of oxides on the particle surfaces which may interfere with compaction and sintering and which, if sufficiently stable, may be retained in the final sintered structure.

Apparent Density

The weight of a unit volume of loose powder is called the "apparent density" of the powder. The technique for measuring the apparent densities of metal powders involves the use of a standard funnel device as detailed in ASTM Standards B212 (free flowing powders; Hall funnel) and B417 (non-free-flowing powders; Carney funnel). The apparent density, usually expressed in units of grams per cubic centimeter, is an extremely important powder characteristic because it provides an estimate of the actual volume to be occupied by a certain mass of loose powder which, in turn, is related to the size (height) of the compaction tooling to be used for a given part and the distance (stroke) the press must move to compact the powder. For a given metal, the apparent density will depend on particle size and size distribution, particle shape and surface topography, particle packing efficiency (determined by size and shape considerations), and internal particle structure (porous or nonporous). Generally the apparent density (expressed as a percentage of theoretical density) of P/M powders will range from about 15% (fine, irregular particles) to about 60% (coarse, smooth spherical particles).

As Hirschhorn[1] has appropriately pointed out, in order to appreciate the general effects of particle size and shape on apparent density, one must remember that space can only be completely occupied by certain regular polyhedral particle shapes, such as cubes, which are all identical in size and perfectly positioned relative to one another. Real powder particles cannot completely fill space. One approach to increasing the apparent density of a powder is to alter its particle size distribution to provide an optimum

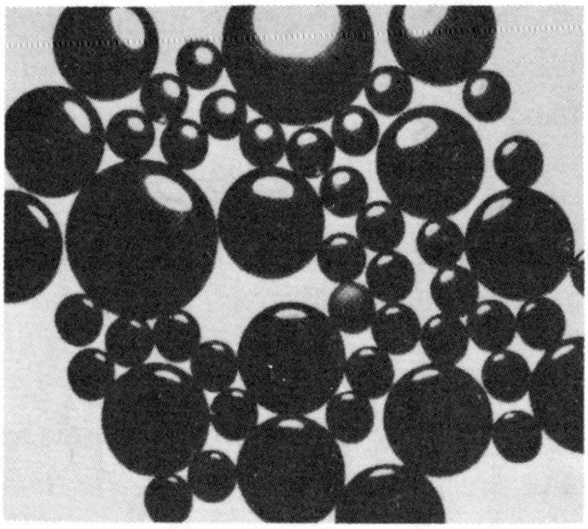

Fig. 8. Idealized illustration of bridging effect created by the presence of fine particles.[1]

number of fine particles to partially fill the voids between larger particles. However, if present in sufficient quantity, fine particles may lead to a reduction in apparent density by decreasing the packing efficiency of the larger particles. This effect, often referred to as "bridging", is shown in Fig. 8. Figure 9 illustrates four basic apparent density variations which might result from addition of fine particles to a coarse distribution. Case 1 depicts the somewhat unusual situation wherein the apparent density increases continuously as fine particles are added, indicating that bridging effects are less dominant than void filling and that the fine particles have a higher apparent density than the coarse material. This could occur in those powders whose finer particles tend to be smoother and more regularly shaped—e.g., water-atomized iron and iron alloy powders. Cases 2 and 3 illustrate relatively normal behavior where bridging becomes a dominant factor if fine particles are added in sufficient numbers. Case 4 represents severe bridging resulting from even very small additions of fine particles, as might be expected if the fine particles were irregularly shaped with very rough surfaces.

Flow Rate

In conventional P/M operations, where automated high-speed compacting presses are routinely used, the ability of a

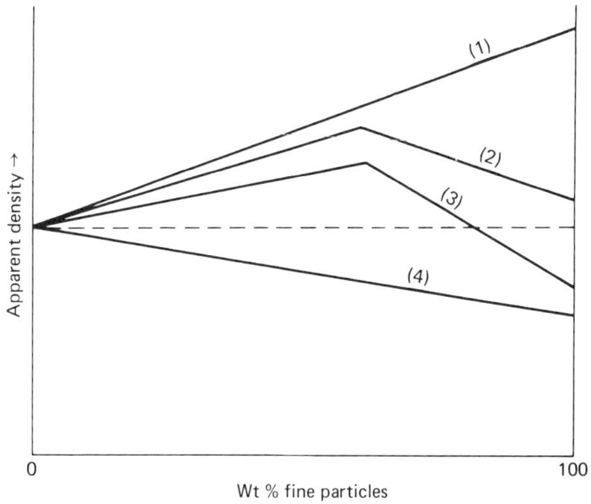

Fig. 9. Possible effects of fine particle additions on the apparent density of an initially coarse particle distribution.[1]

metal powder to flow rapidly and smoothly from a container into the compaction tooling cavities is extremely important. This characteristic of powders is, like apparent density, primarily related to interparticle friction. In general, those factors which reduce interparticle friction and/or increase apparent density (such as particle smoothness and sphericity) will improve the flow behavior of a powder.

The terms "flow rate", "flow time" and "flowability" are often used virtually interchangeably in discussions of powder flow. The standardized test for the flow of metal powders (ASTM B213) actually measures the time required for 50 grams of the loose powder to flow under the action of gravity through the orifice of a Hall funnel. Of course, the mass flow rate of the powder is inversely proportional to this flow time.

Compressibility

In general terms, the compressibility (or compactibility) of a powder is a measure of the amount of densification which can be achieved by applying pressure to the powder mass. ASTM Standard B331 prescribes the general method used for this evaluation, which involves uniaxial compaction in a rigid die. Typically, the compressibility of the powder is reported as the density after compaction at 30 tons per square inch (415 MPa). Alternatively, the compacted densities at several pressures may be of interest, resulting in a compressibility curve similar to those illustrated in Fig. 10. It should be noted here that the term "green" density, ρ_G, is simply the density of the compacted powder measured after ejection from the die cavity.

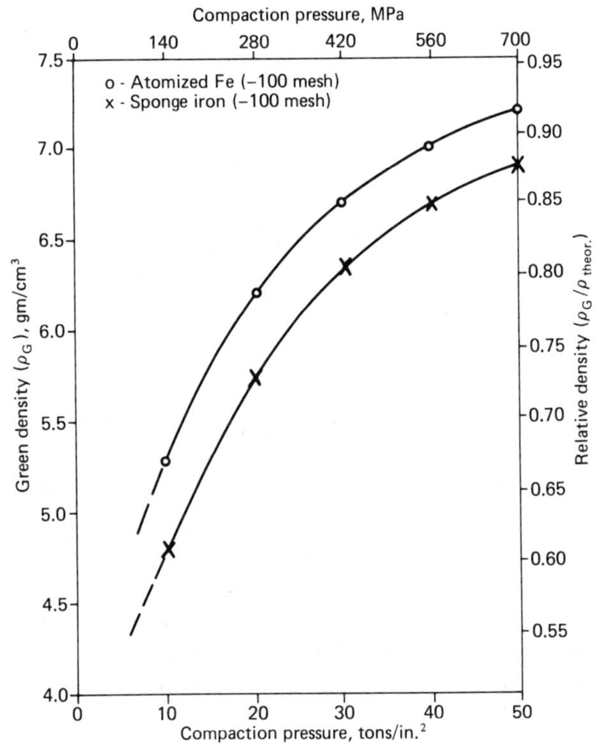

Fig. 10. Compressibility curves for atomized and sponge iron powders. Die wall lubrication used.

Like its apparent density and flow rate, a powder's compressibility will be influenced considerably by the effects of interparticle friction as well as by the integrity of the particle itself (presence or absence of internal particle porosity). For the curves in Fig. 10, the atomized iron powder is somewhat more regular in shape, is smoother, and has less intraparticle porosity than the sponge iron powder. Consequently, by comparison, the atomized material is considered to be a high-compressibility powder. In addition to the factors of shape and internal structure, compressibility may be greatly affected by the ease with which the particles can deform during the compaction process. This point will be briefly considered in the next section.

Compaction of Powders

Having introduced the concept of a powder's compressibility, it is appropriate that we next examine the P/M compaction process more carefully since it is a basic and very important unit operation. In addition to consolidating a powder into some intended shape, compaction serves to provide the desired amount and type of porosity along with adequate strength ("green" strength) for subsequent handling prior to sintering.

Preliminarily it should be noted that prior to compaction most P/M powders are blended with a particulate lubricant such as zinc stearate or lithium stearate in amounts of a few volume percent. This is necessary to reduce friction between the powder and the tooling surfaces (die cavity, punches, core rods), permitting higher effective pressures to be transmitted to the powder mass and thus yielding a higher compacted (green) density for a given applied pressure. The lubricant also aids significantly in permitting the compact to be ejected from the die cavity at relatively low loads and greatly reduces wear on the tooling surfaces. Because the lubricant inside the powder compact serves little purpose and, in fact, is generally detrimental to compact quality and inconvenient in subsequent sintering operations, it would be desirable to apply the lubricant only where it is needed—i.e., on the tooling surfaces.

Unfortunately, a totally satisfactory high-speed lubrication technique has not yet been developed, and most powders are compacted with the lubricant "admixed".

Compaction Techniques

There are several ways to compact powders. P/M processing may involve compaction in rigid dies, ambient or high-temperature isostatic compaction in flexible tooling, high-energy-rate forming (HERF), or consolidation by forging, by rolling or extrusion, or by vibratory packing. By far, the largest number of P/M components are compacted in rigid dies using mechanical or hydraulic presses. The other techniques listed above are both important and interesting, but space does not permit us to examine them now. Because rigid-die compaction presents a sufficient opportunity to examine some of the general ideas of compaction, we will concentrate on this technique alone, and all references to compaction throughout the remainder of this chapter will imply the use of this technique.

There is great variety in the tooling designs and press motion sequences which are employed in rigid-die compaction. These range from very simple rapid systems for producing single-level parts of the type shown in Fig. 11(a) to the much more complex systems used for compacting the multilevel parts depicted in Fig. 11(b). A schematic look at the basic steps of powder compaction using rigid tooling should be sufficient here to provide the reader with a basic understanding of the process. Figure 12 illustrates the steps involved in compacting simple, relatively thin parts (e.g., Class I) by the downward motion of only the top punch, with the die barrel and bottom punch remaining stationary. The compacted part is ejected from the die by an upward movement of the bottom punch with the top punch removed. This method is commonly called "single-action" compaction. Thicker or somewhat more complex parts can be produced by a modification of this procedure in which both the upper and lower punches move against the powder mass to compact it. This "double-action" process is illustrated in Fig. 13. A variation of the double-action method involves

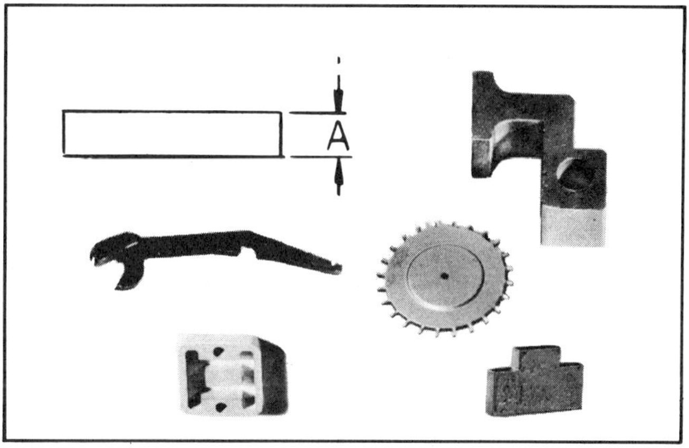

(a)

(b)

(a) Class I: thin, single-level parts (A < 0.250 in.). (b) Class IV: multilevel parts of any thickness or contour.

Fig. 11. Two classes of P/M parts compacted in rigid dies. [11]

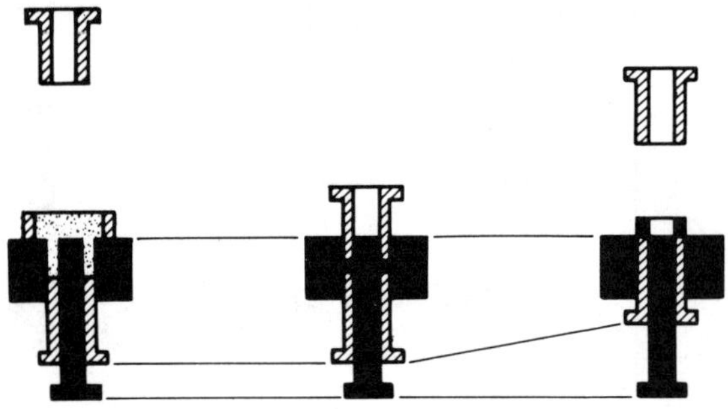

Fig. 12. Basic steps in single-action powder compaction in a rigid die. [11]

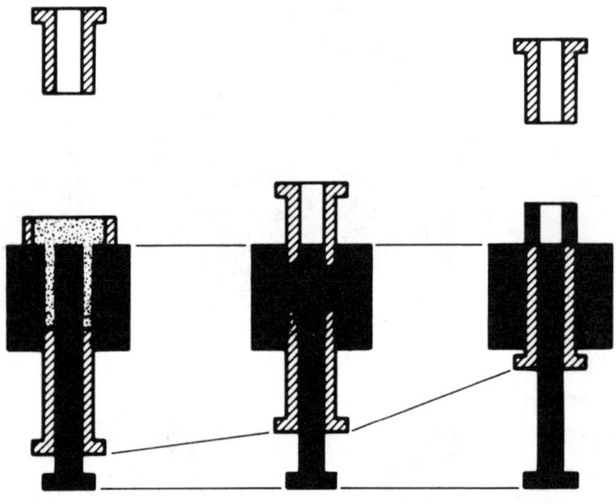

Fig. 13. Basic steps in double-action powder compaction in rigid tooling. [11]

the use of a die barrel which is only modestly supported against the downward movement of the top punch. This "floating-action" procedure is shown in Fig. 14. As the friction builds during the downward movement of the top punch, the die barrel also begins to move downward over

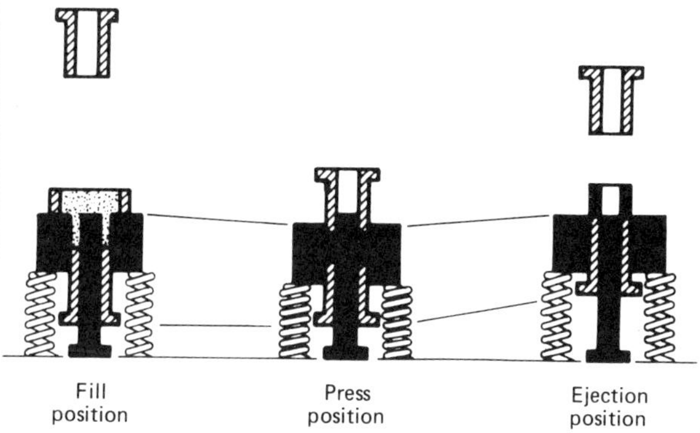

Fig. 14. Floating-action die compaction system.[11]

the stationary bottom punch, effectively simulating compaction from both ends.

Other more complex compaction tooling systems are used regularly in the P/M industry. Most of these, including the "multiple motion", "reflex" and "withdrawal" designs, are adequately discussed elsewhere[1,2,11] and will not be considered here.

Powder Behavior During Compaction

The actual behavior of powder particles under the action of an applied compacting force is of fundamental importance to the P/M process. However, because of the physical and geometrical complexities associated with the powder mass, little detailed knowledge is available to provide more than the simplest of qualitative insights into the phenomena involved. In 1946, Seelig and Wulff[12] suggested that three general events take place in the powder during compaction:

- The particles undergo bulk movement such that they are rearranged into a more efficiently packed array.
- Both elastic and plastic deformation are imparted to the particles.
- The particles may fracture (fragment) as the compacting load is increased.

The rearrangement of the particles is limited by interparticle friction as well as by the friction between the powder and the compaction tooling. Such bulk particle movement generally occurs at relatively low compaction pressures and accounts for most of the early densification in the compact.

When the probability of additional bulk particle rearrangement is sufficiently low, substantial particle deformation may take place and result in further densification (reduction of the porosity level) of the powder mass. This is the major densification mechanism involved in the attainment of high green densities and may also lead to the significant additional interparticle contact area essential to high green strength. The extent of plastic deformation depends on the ductility, yield strength and strain hardening characteristics of the constituent particles. Deformation is probably of relatively little importance in the compaction of the harder powders such as tungsten, various carbides, and ceramics such as alumina. Ultimately, the third stage involving particle fracture is encountered. Such particle fragmentation leads to still further densification by virtue of the fact that some of the newly formed small pieces can move into the remaining pore spaces between the larger particles.

The nature of compressibility behavior (such as that shown in Fig. 10) has been studied extensively. This work and the various resulting empirical relationships between green density and compaction pressure have been reviewed by Bockstiegel and Hewing[(13)], and the interested reader is referred to this paper.

The configuration of the porosity in a compact is a complex function of many variables related to the powder and to the details of the compaction process itself. For most powders, there exists a significant range of compaction pressures which may be used without losing the interconnected nature of the porosity. In general, pressure can be an effective process variable for controlling the ultimate level of interconnected porosity for P/M applications such as filters or parts requiring subsequent infiltration with liquid metals or impregnation with a variety of nonmetallic materials.

Green density can be used as a measure of the effectiveness of the compaction operation. Ordinarily, we consider a single measured over-all value for this density. However, because a mass of powder does not behave as a fluid—i.e., is not acted upon uniformly by the applied compaction force—there may be large variations in density within a compact if the compaction process is not selected wisely. These density variations result from stress distributions in the compact promoted by frictional forces at the tooling/powder interfaces. When powder is compacted in rigid tooling, particle movement is generally in the pressing direction. The green density is greatest near the moving punch, and decreases in regions remote from the punch, particularly near the die wall. An actual green density distribution in a nickel powder compact pressed at about 100,000 psi (690 MPa) to a length:diameter (L/D) ratio of 0.87 is shown in Fig. 15. Here, single-action pressing was employed from the top, and the density distribution is typical for this mode of compaction of a

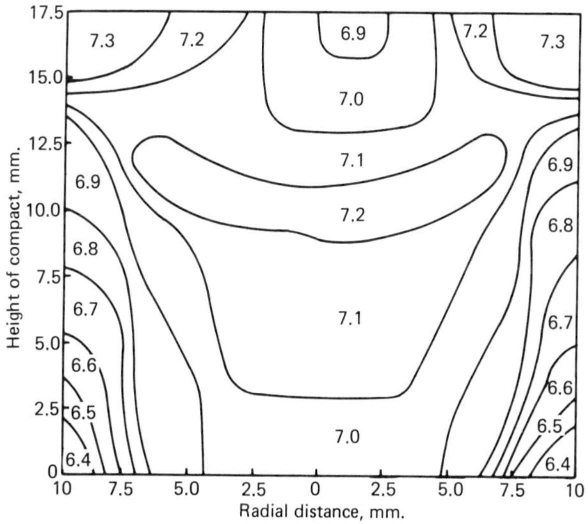

Fig. 15. Green density distribution in a nickel powder compact pressed at about 100,000 psi (690 MPa) from the top only.[14] Density values, in g/cm^3, inferred from macrohardness measurements.

relatively thick part. At the cylindrical surface, the density of the compact decreases substantially from top to bottom. It should be noted that the L/D ratio of this compact (0.87) is quite large and that the observed density gradients would be somewhat different and certainly much less severe if double-action compaction were employed.

Strength of Powder Compacts

The strength of a powder compact, usually called its "green strength," is promoted largely by interparticle cohesion resulting from initial particle surface asperities and may be significantly enhanced by the plastic deformation which occurs during compaction of ductile powders. The green strength of a compact generally can be increased[1] by:

- Using fine powders
- Using powders with irregular particle shapes and rough surfaces
- Increasing the green density (using higher compaction pressure)
- Decreasing the surface contamination on the particles
- Decreasing the amount of admixed lubricant or other powder additives.

The significance of some of these variables in determining the green strength of iron powder compacts can be appreciated by examining the data presented in Fig. 16. The sponge iron powder has about the same particle size distribution as the atomized material. However, the sponge iron particles are more irregularly shaped and have somewhat rougher surfaces and more intraparticle porosity. These characteristics are reflected in the substantially higher green strength values for this material, especially at higher compaction pressures (higher green densities). The addition of about 1 wt % zinc stearate to the sponge powder reduces the green strength by almost 40% at a green density of 6.3 g/cc (80% of theoretical density).

The green strength of a compacted powder is usually measured by a three-point bend test, as described in ASTM Standard B312. The strength of the compact is

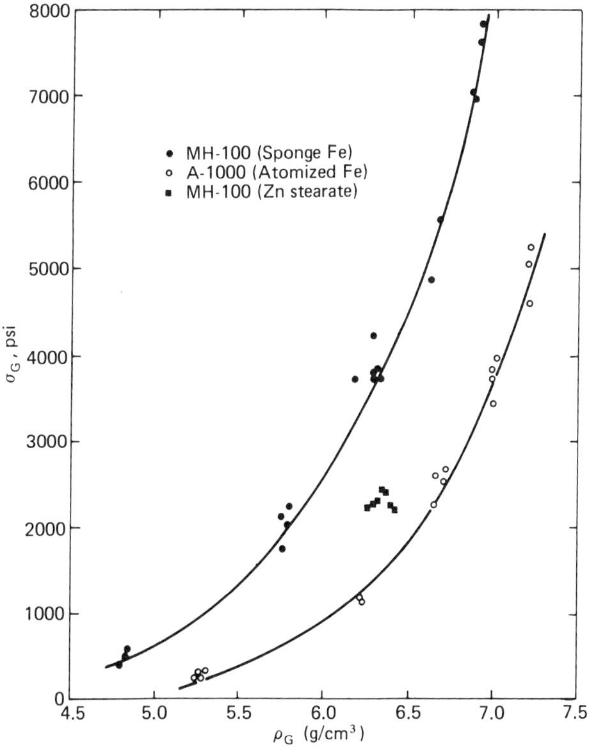

Fig. 16. Green strength vs green density for MH-100 sponge iron and A-1000 atomized iron powders. Die wall lubrication used (●, ○). 1 wt % zinc stearate admixed in MH-100 (Zn stearate) specimens (■).

governed by its weakest structural feature—the interparticle surfaces—and not by the strength of the actual solid material. A normal compact will simply fracture in a brittle manner when subjected to sufficient stress.

Sintering

For purposes of establishing a general definition, conventional sintering may be considered to be the bonding of adjacent surfaces in a mass of powder, or a compact, by heating (ASTM B243). Sintering is a physically complex phenomenon which results in numerous structural changes within a powder mass and significant improvement of its mechanical properties. Densification (removal of porosity) is often associated with sintering. However, in commercial practice where close dimensional control is required, more than a minimal amount of densification is often undesirable and the sintering process must be controlled accordingly.

Conventional sintering involves heating of the green compact to some temperature below its melting point or solidus in a protective or reducing atmosphere at normal pressure. Variations include sintering in the presence of transient or stable liquid phases and hot pressing (simultaneous compaction and sintering). Both hot pressing and liquid phase sintering are of significant commercial importance, but the discussion here will focus only on an introduction to conventional solid-state sintering.

In general, the following changes occur in a powder compact when it is subjected to conventional sintering:

- Particles begin to bond together, increasing the strength of the compact as well as its electrical and thermal conductivity.
- As sintering time is extended (or sintering temperature raised), the strength continues to increase.
- The increase in strength is often accompanied by a decrease in the amount of porosity (increase in compact density).
- Grain growth may occur, and the final grain size may be much larger than the original powder particle size.
- Pores become smoother and more spherical in shape as sintering progresses.
- In a suitable sintering atmosphere, adsorbed gases are removed from the compact and particle surface oxides are reduced.

Development of Mechanical Properties

Typically, as sintering progresses, strength and ductility of the compact both increase. These increases are related to an increase in interparticle bonding and to changes in the amount and morphology of the porosity. The data of Squire[15] shown in Fig. 17 illustrate quite clearly this

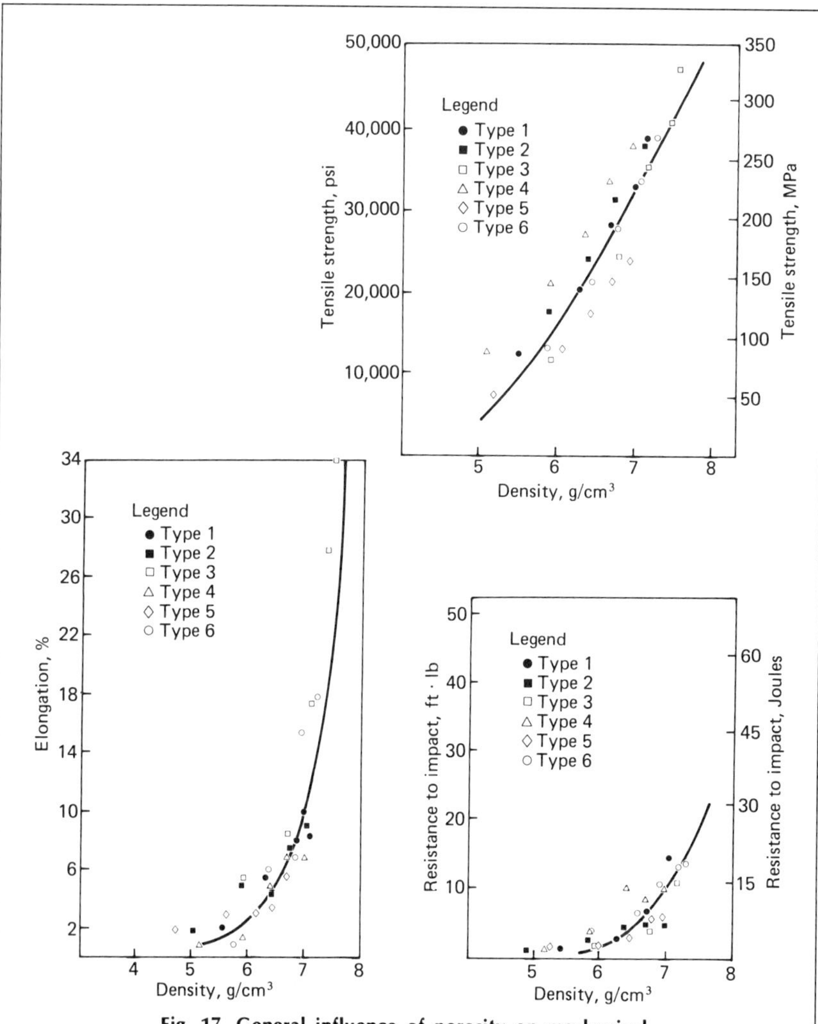

Fig. 17. General influence of porosity on mechanical properties of sintered compacts made from six different types of iron powder. Compacts were sintered 1 h at 1100°C. (Data from Ref. 15. Figures reproduced from Ref. 2)

potentially dramatic effect of porosity on mechanical properties. It is very apparent that porosity is detrimental to strength- and ductility-related properties, the effect being most pronounced on tensile elongation and impact resistance as density approaches the theoretical value (7.87 g/cm^3 for iron). Pores in the structure provide sites for stress concentration and strain localization, thus allowing deformation and fracture to occur at relatively low levels of applied load. Smoothing, spheroidization and isolation of porosity are beneficial to strength, ductility and toughness. Some of these points are discussed in the paper by Vedula and Heckel[16]. These structural changes, along with pore shrinkage, generally may be considered to occur late in the sintering cycle.

The mechanical properties of P/M parts containing a few volume percent of porosity are typically inferior to those of their full-density counterparts produced by casting and subsequent deformation processing. However, full-density P/M materials produced by preform forging, hot pressing, etc., often exhibit properties which equal or exceed those of competitive materials. In part, this is due to the inherently finer grain structure and relative freedom from nonmetallic inclusions which can result from P/M processing.

Development of Microstructure

Except for the presence of porosity and the general absence of preferred crystallographic orientation (annealing texture), sintered microstructures are similar to those of equivalent wrought and annealed materials. It is convenient to think of sintering in terms of several distinct stages of microstructural change. Although opinions vary somewhat concerning the exact events constituting each stage, it is important to realize that this description is at best very qualitative and that these "stages" may occur virtually simultaneously within the compact. Nevertheless, the concept does help to introduce the physical nature of the sintering process. In Hirschhorn's discussion of this subject[1], sintering is divided into six stages:

1 Interparticle bonding
2 Neck growth
3 Closure of pore channels
4 Rounding of pores
5 Pore shrinkage (densification)
6 Pore coarsening.

Interparticle Bonding. Transport of atoms at particle contact points leads to the establishment of physical bonding and grain boundaries at these points. These initial bonds are developed very rapidly—even as the compact is being heated to temperature.

Neck Growth. Continuing mass transport leads to the development of distinct "necks" between particles, growing from the initial bonds. This stage greatly increases the strength of the compact but does not, itself, involve any densification. Neck growth between coarse atomized 316L stainless steel particles is shown in Fig. 18.

Pore Channel Closure. Continued neck growth can cause pore channels within the compact to be closed, leading to isolated porosity. It is very important to control this event in sintering powder metallurgy products which

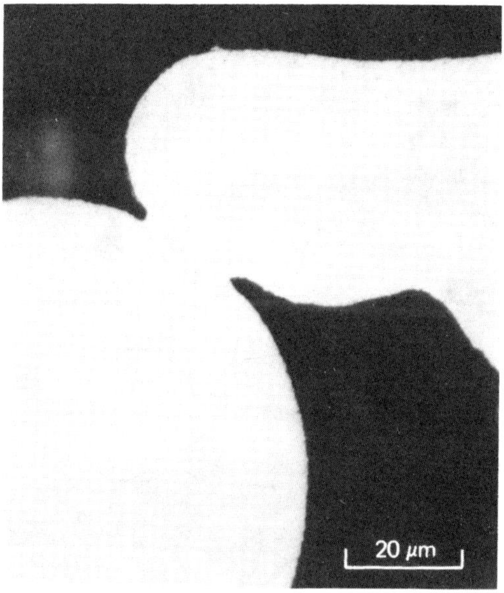

Fig. 18. Photomicrograph showing substantial neck growth between two particles of 316L stainless steel sintered 3 h at 1300°C.

utilize interconnected porosity—for example, filters, self-lubricating bearings, etc.

Pore Rounding. As neck growth reaches an advanced stage, the material transported from the general particle surfaces to the neck regions results in a smoothing of the pore walls (reduction of surface area). This can be an important structural change with respect to improving the ductility and toughness of the sintered compact. This stage of sintering is illustrated by the idealized three-sphere models in Fig. 19. A photomicrograph of an iron compact sintered to an extent where substantial pore rounding has occurred is shown in Fig. 20.

Pore Shrinkage. As sintering progresses, the pores in the compact may begin to shrink in size and decrease in number, resulting in densification. This stage must involve

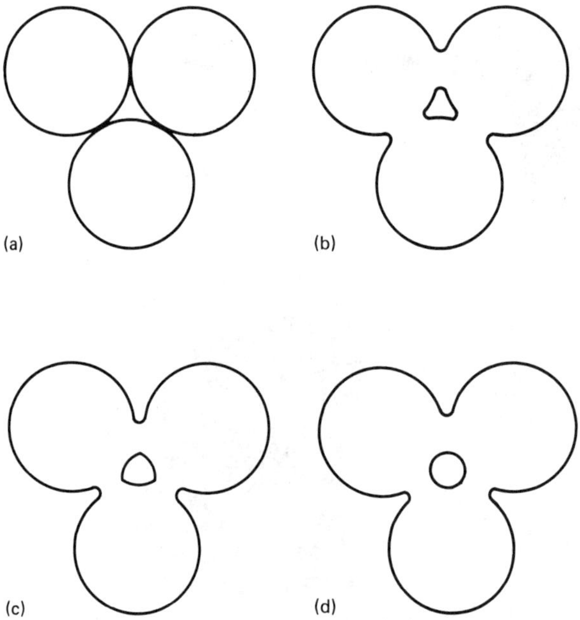

Fig. 19. Three-sphere sintering models.[1] (a) Original point contacts. (b) Neck growth. (c) and (d) Pore rounding.

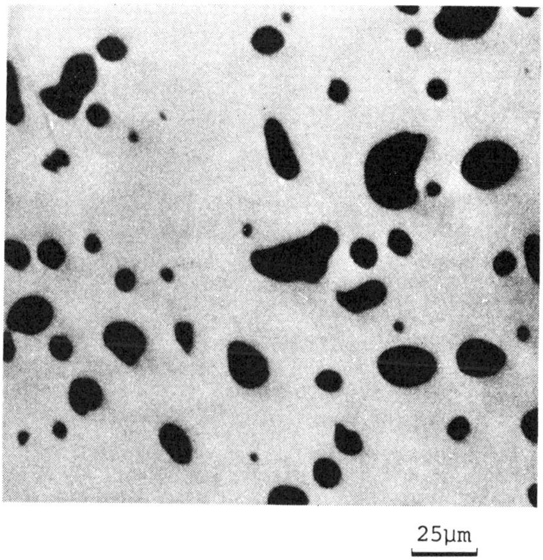

Fig. 20. Photomicrograph of −400 mesh atomized iron compact (40 tsi) sintered 100 h at 1150 °C in hydrogen.[16] Specimen unetched. Note extensive rounding of the porosity.

extensive diffusion and annihilation of vacancies. Not all porosity can be eliminated during this stage of sintering: some residual pores will be left (particularly inside of grains) even after very long sintering times. An example of a sintered structure in which significant pore shrinkage has occurred is shown in Fig. 21. This material is molybdenum (powder shown in Fig. 3) whose green density after 60,000-psi isostatic compaction was 70% of theoretical. After sintering at 1850°C (3362°F) for nine hours, the density increased to 98% of its theoretical density. Note that most of the residual porosity is in the interior of grains and that the grain size is about 30 μm, which is an order of magnitude larger than the average particle size (3 μm) of the starting powder.

Pore Coarsening. This stage is not considered to be very important in most commercial sintering operations. When it occurs, it involves an increase in the size of some of the

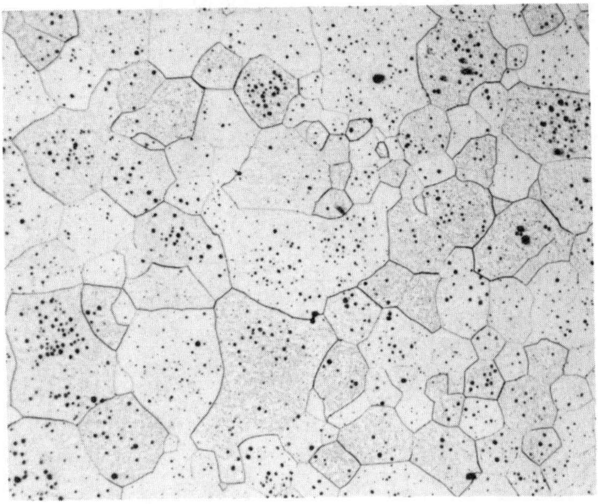

Fig. 21. Microstructure of sintered molybdenum. Density, 98% of theoretical. Average grain size, approximately 30 μm. Magnification, 250× (shown here at 75%).

larger pores and concurrent elimination of some smaller pores. No net change in the pore volume fraction is involved, so the average distance between pores is increased. In many respects, this microstructural change is analogous to the precipitate coarsening which occurs during severe overaging of conventional age-hardening alloys.

Sintering Mechanisms

The various structural changes that take place during sintering obviously depend on mass transport to and from various regions within the powder compact. A net decrease in free energy must be associated with sintering if these structural changes are to occur spontaneously. In a simple material, this decrease in free energy is associated with a decrease in the total surface area (and therefore surface energy) of the compact. All of the basic structural changes, such as neck growth, pore rounding, etc., lead to reduced total surface area within the compact. Thus it is not surprising to find that the greater the original surface area the larger the driving force for sintering—i.e., compacts of fine powders sinter faster than compacts of coarse powders.

Probably the most dominant mass transport mechanism involved in normal sintering is solid-state diffusion through the crystalline lattice and via short-circuit paths such as particle surfaces and grain boundaries. In metals with very high vapor pressures, such as zinc and cadmium, evaporation of material from the convex particle surfaces and subsequent condensation of this material at the concave necks between particles is a probable transport mechanism. However, this mechanism is not considered to be particularly active in the sintering of most P/M materials, because their vapor pressures are quite low at normal sintering temperatures. It is also possible, due to effects of surface tension at the interparticle necks, to promote some mass transport by self-induced plastic deformation early in the sintering cycle, when the radii of curvature of the necks are very small.

In considering the nature of diffusion during sintering, it is perhaps easiest to think in terms of the motion of lattice vacancies in the material, bearing in mind that atoms move opposite to the vacancies. Using a two-sphere model to illustrate the concept of vacancy diffusion (Fig. 22), we can imagine that the spheres are crystalline powder particles containing unoccupied lattice sites (vacancies) whose concentration will vary with location in this model. Due to surface-tension effects related to the curvatures of the neck surfaces (radius of curvature, ρ) and particle surfaces (radius of curvature, r), the Gibbs-Thomson equation predicts a chemical potential gradient arising from a vacancy concentration in the neck which is greater than the vacancy concentration in the particle away from the neck. So, there is a vacancy concentration gradient which provides the driving force for the movement of atoms into the neck region and the corresponding movement of vacancies away from the neck. These same considerations apply to the shrinkage of pores—the vacancy concentration is high near pores, leading to mass transport into the pores and subsequent vacancy annihilation (densification).

The rate at which diffusion occurs in a given structure depends primarily on temperature. Diffusivity increases

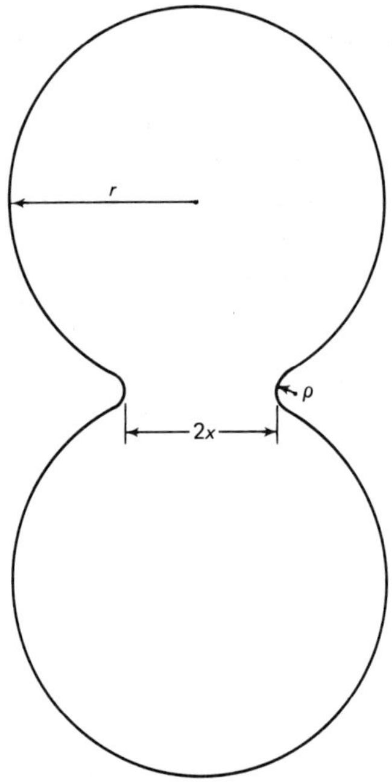

Fig. 22. Schematic illustration of two spherical particles sintering together.[2]

exponentially with temperature. Therefore, sintering rates are strongly influenced by the temperature at which the process is conducted.

Alloying by P/M Techniques

The sintering operation can be utilized to produce alloys from compacted mixtures of two or more elemental and/or masteralloy powders. Examples include:

- Sintering of iron-graphite blends to produce carbon steel (see Fig. 23)
- Sintering of copper-tin blends to produce P/M bronzes utilizing a transient liquid phase during sintering
- Sintering of titanium-plus-masteralloy blends to make alloys such as Ti-6Al-4V

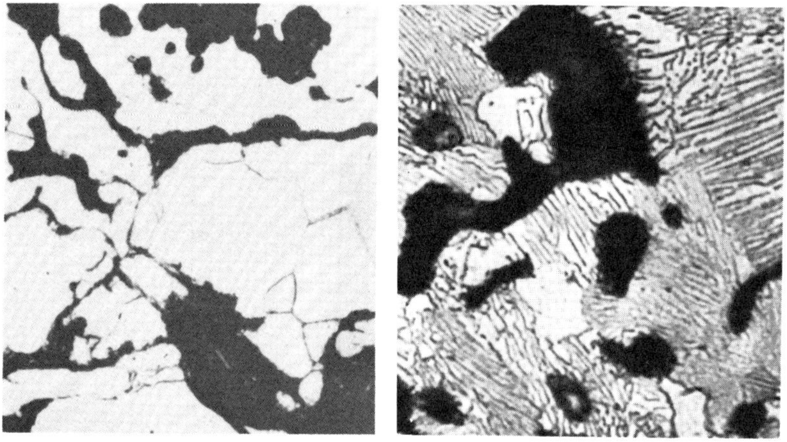

(a) (b)

(a) As-compacted, with no bonding or interdiffusion. (b) Pearlitic structure resulting from slow cooling of sintered compact (120 min at 1200°C) in which carbon was uniformly dissolved in austenite. Note rounded porosity in (b) compared to (a). Magnification (both), 800×.

Fig. 23. Iron-graphite powder mixture compacted to a green density of 6.1 g/cm^3 (Ref. 17).

- Sintering of blends of tungsten and rhenium powders to produce alloys for special filament-wire applications (see Fig. 24).

The development of a homogeneous microstructure during sintering is accomplished by interdiffusion of the component elements. In the simplest sense, the driving force for this interdiffusion is related to the concentration gradients in the compact, and the process involves the progressive reduction of these gradients with time at the sintering temperature. This technique for producing P/M alloys is often called "homogenization processing"; the physical aspects of this technology have been eloquently discussed by Heckel(19). The path by which homogenization occurs depends on the nature of the phase diagram

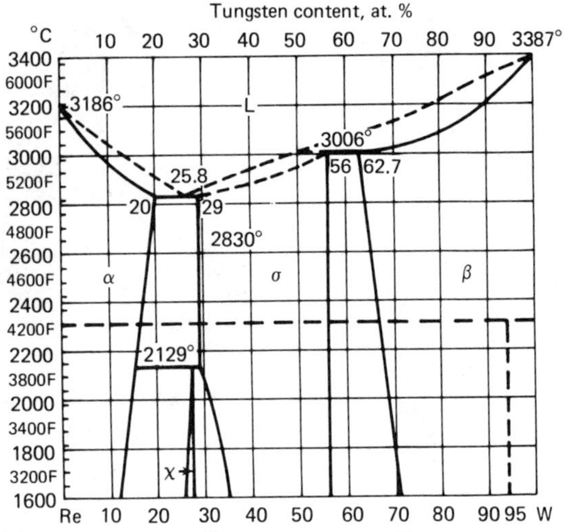

(a)

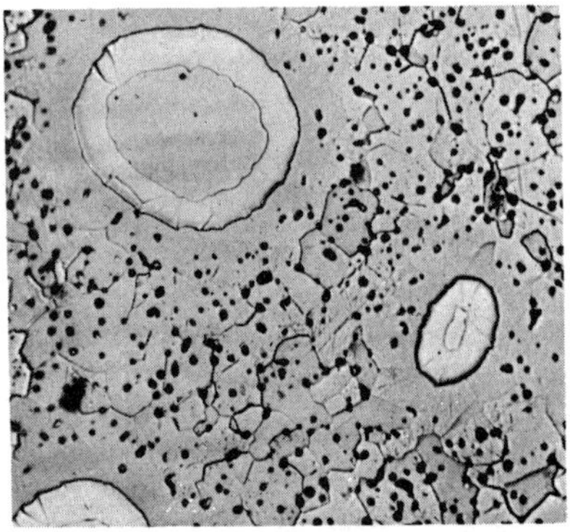

(b)

(a) Re-W phase diagram.[18] (b) Photomicrograph (at 500×) of a 95W-5Re powder compact sintered for 10 min at 2325°C.[17] Porous matrix is tungsten-rich solid solution (varying compositions of β). Circular features are intermediate σ phase surrounding rhenium-rich solid solution (α). Complete homogenization would result in a single-phase (β) microstructure.

Fig. 24. Rhenium-tungsten phase diagram and photomicrograph of a W-Re powder compact.

of the alloy system in question at the sintering temperature. As illustrated in Fig. 24 for the tungsten-rhenium binary system, homogenization may involve the formation and subsequent dissolution of one or more intermediate phases. Also, either stable or transient liquid phases may form as a result of interdiffusion in systems such as copper-tin where the sintering temperature is abovc the minimum solidus temperature of the alloy system.

Acknowledgments

The author extends his sincere appreciation to R. Farrell and L. Sutter of Michigan Technological University, Dr. E. Klar of Glidden Metals Division of SCM Corporation and Dr. J. E. Smugeresky of Sandia National Laboratories (Livermore, CA) for their assistance in providing some of the illustrations used in this presentation. The cooperation of the American Powder Metallurgy Institute and the Metal Powder Industries Federation is also gratefully acknowledged.

References

1. J. S. Hirschhorn, *Introduction to Powder Metallurgy*, Amer. Powder Met. Inst., Princeton, NJ, 1969
2. F. V. Lenel, *Powder Metallurgy Principles and Applications*, Metal Powder Ind. Fed., Princeton, NJ, 1980
3. P. Gummeson, *Powder Metall.*, *15*, 1972, p 67-94
4. E. Klar and W. Shafer, "High Pressure Gas Atomization of Metals", in *Powder Metallurgy for High Performance Applications*, J. Burke and V. Weiss (eds.), Syracuse, NY, 1972, p 57-68
5. D. W. Smith, "Powder Characterization", *Encyclopedia of Materials Science and Engineering*, Pergamon Press, New York (to be published in 1983)
6. T. Allen, *Particle Size Measurement*, Chapman and Hall, London, 1968
7. R. Irani and C. Callis, *Particle Size: Measurement, Interpretation and Application*, Wiley, New York, 1963
8. C. Orr and J. Dallavalle, *Fine Particle Measurement*, Macmillan, New York, 1959
9. H. Hausner, *Handbook of Powder Metallurgy*, Chemical Publ. Co., New York, 1973

10. J. Beddow and T. Meloy (eds.), *Testing and Characterization of Powders and Fine Particles,* Heyden and Son Publ., London, 1980
11. *Powder Metallurgy Equipment Manual*, Metal Powder Ind. Fed., Princeton, NJ, 1977
12. R. Seelig and J. Wulff, *Trans. AIME*, *166*, 1946, p 492-504
13. G. Bockstiegel and J. Hewing, *Arch. Eisenhüttenwesen*, *36*, 1965, p 751-767
14. G. Kuczynski and I. Zaplatynskyj, *Trans. AIME*, *206*, 1956, p 215
15. A. Squire, *Trans. AIME*, *171*, 1947, p 485-505
16. K. Vedula and R. Heckel, *Modern Dev. in P/M*, *12*, 1981, p 759-777
17. *Metals Handbook* (8th Ed.), *7*, ASM, 1972
18. *Metals Handbook* (8th Ed.), *8*, ASM, 1973
19. R. W. Heckel, "Diffusional Homogenization of Compacted Blends of Powders", in *Powder Metallurgy Processing*, H. Kuhn and A. Lawley (eds.), Academic Press, New York, 1978, p 51-97

Chapter 3

Designing for Powder Metals

Ron Sadlon
Powder Metal Products Division
Remington Arms Company, Inc.

Powder metallurgy offers designers and engineers a method of obtaining components possessing a wide range of characteristics in a variety of compositions at reasonable cost. If maximum benefits of P/M are to be realized, it is necessary that the basics of the process be understood and that parts be of proper design. Designs must be related to method of manufacture in order to capitalize on the design advantages unique to the P/M process while allowing for its limitations.

Close cooperation between supplier and user is required in the initial stages of planning and design. Whenever possible, a powder metal part should be designed in close conjunction with the design of the entire assembly. Small modifications in the assembly sometimes will permit a better and lower-cost powder metal part to be produced.

The basic design parameters of powder metal products will be reviewed in this chapter. Many applications require deviations from these basics, and those concessions that must be accounted for will be discussed.

Pressing Equipment

Whether or not a structural part can be made successfully by powder metallurgy depends largely on the success of the briquetting or pressing operation. Equipment used for pressing of metal powders is similar to that used in the ceramic, carbide, ferrite and plastic industries. Specialty presses include anvil, rotary and isostatic presses.

Anvil pressing is the simplest and most economical high-speed method of compacting powders. It is especially well suited for parts with one flat side, because the part is compacted against an anvil by the upward action of a lower punch contained in a die. Parts are limited to simple shapes and relatively small sizes (less than 1 $in.^2$ in cross-sectional area; less than 0.5 in. thick).

Rotary presses are used for high-volume production of parts of simple shape. Equipment with 40 sets of tools and the ability to run 5000 compacts per minute is not uncommon. These presses, commonly referred to as "pill" or "tablet" presses, are used in the pharmaceutical industry.

In isostatic presses, compaction takes place in a sealed pressure chamber by exertion of fluid pressure on an elastic container pre-filled with powder. Complex shapes can be produced that are not obtainable using standard powder metal pressing techniques. Generally, cycle rates are much lower than those of conventional techniques.

The majority of powder metal presses fall into two basic classes: mechanical and hydraulic. Both have advantages and disadvantages. The material to be compacted, the required total pressure, the shape of the part, the required output and other factors all influence the decision as to which type of press is best suited to a given application. Mechanical presses generally are advantageous for high-output applications but have a higher rate of wear and are limited in their scope of application. The initial cost of a mechanical press is directly proportional to its size, whereas the increase in cost of hydraulic presses with size tends to taper off at the higher levels due to the use of common controls on all systems. Labor-reducing devices (robots or transfer systems) further decrease the potential savings advantages of mechanical presses.

Transferal of tooling between parts producers usually is not recommended. Dimensional discrepancies can result from the use of different processing equipment and techniques. Tool adaptation can be as costly as new fabrication of tools, because most press manufacturers supply equipment to take adapters of each individual customer's design.

Compaction of Powders

As stated above, successful production of structural parts by P/M techniques depends largely on the success of the compacting operation. Metal powders do not flow freely; friction between adjacent particles, and between particles and die walls, prevents flow—especially in long, narrow cavities. Therefore, unless provision is made for proper powder distribution, the volume of powder in the die may be insufficient in some areas.

Because there is practically no lateral flow in the die, the proper quantities of powder must be placed exactly where they are required in order to obtain uniform density. This requirement is particularly essential in producing parts with thin walls or variable cross sections. The fact that powders do not flow laterally also places some limitations on the contours that can be produced in the pressing operation.

The last rule of powder metallurgy that must be considered is that loose powder occupies approximately twice as much volume as that occupied by the final pressed part.

Tooling

In the pressing cycle, a controlled amount of blended powder is fed automatically into a precision die and is pressed to form a compact with the size and shape of the finished part. A typical set of tools for producing a cylindrical bearing is shown in Fig. 1. Punch members form the top and bottom contours of the cylinder while the die and core rods dictate the external and internal shapes.

Production of a part by use of these tools proceeds in the following sequence (see Fig. 2):

1 The empty die cavity is filled with blended powders, the fill depth being approximately twice the final part thickness.
2 Top and bottom punches simultaneously press the metal powder in the die.
3 The top punch is withdrawn and the lower punch ejects the part from the pressing area and is ready to begin the sequence again.

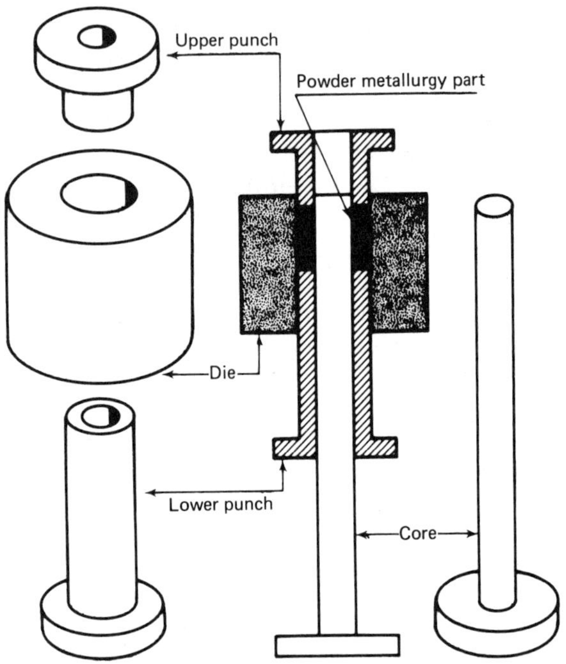

Fig. 1. Typical tooling for production of a cylindrical bearing.

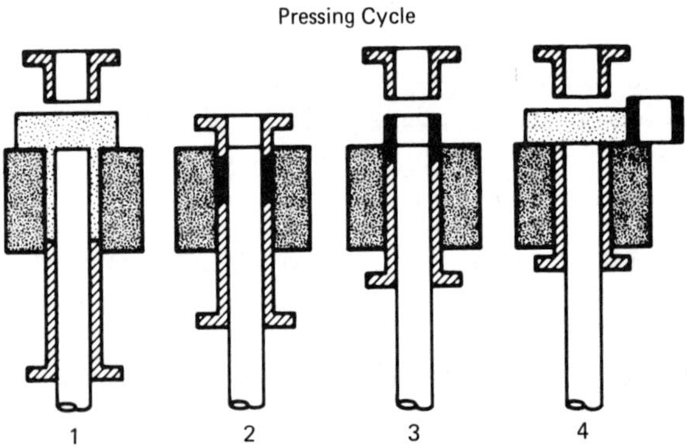

Fig. 2. Sequence of operations for production of a cylindrical bearing.

This tooling concept can be expanded to produce a 1-in.-thick bushing with a 3/8-in. flange, as follows (see Fig. 3):

1 The cavity is filled with 3/4 in. of powder in the flange area and 2 in. in the hub area.
2 The upper punch moves 1/2 in. into the die, the lower inner raises 1/2 in. and the lower outer pushes back 1/8 in.
3 In ejection, both lower punches maintain their support until the flange is above the die; the flange punch movement ceases and the hub punch ejects the hub from the flange member.

This requirement of controlled, independent, supported motion of each tool member limits a part to three or four different major thicknesses.

Design of Powder Metal Parts

Part Size and Shape

Maximum sizes of powder metal parts are basically limited by available equipment and powder characteristics. Conventionally pressed powder metal parts range from

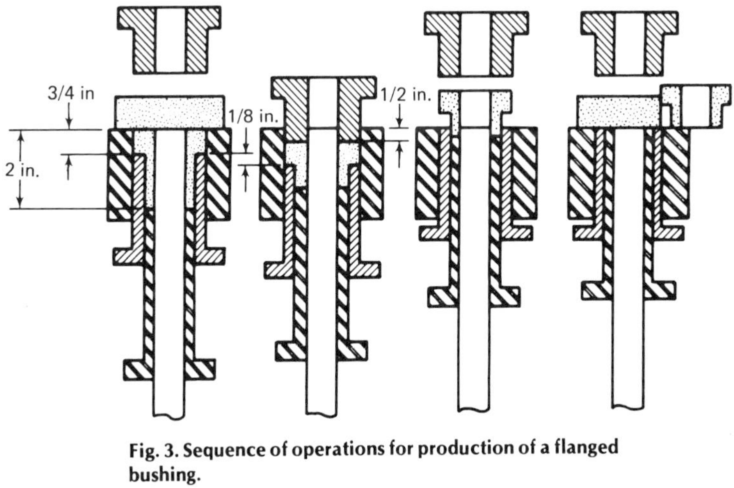

Fig. 3. Sequence of operations for production of a flanged bushing.

about 1/8 to 25 in.^{2} in projected area and from 1/32 to 6 in. in length. The vast majority are less than 15 in.^{2} in area and 2 in. in length. Considering that it takes approximately 40 tons per square inch to compact ferrous powder, a 25-in.^{2} part would require a 1000-ton press. Secondly, because the powder fill is twice as thick as the final part, a 6-in.-thick part requires 12 in. of powder fill. Compaction of powders to these sizes results in extreme density gradients among top, center and bottom, with the center being very low in density (and related properties).

The P/M process offers an unlimited variety of shapes in the die-produced contour of the part. Gears and cams, which would be difficult to machine, are ideal powder metal applications. Square holes or holes of irregular shape present no more difficulty than round holes. Limitations on hole size derive basically from limitations on tool strength. Length-to-diameter ratios of holes should be limited to 4:1. Holes as small as 0.015 to 0.020 in. in diameter have been successfully pressed, but tool life is limited in such applications. Stepped holes may be incorporated to reduce the actual length of the small diameters in the application. Blind holes can be molded to 90% of the thickness of the part from the lower tool members, but much less from the upper members. Tapered holes that would be expensive to machine are readily produced in powder metal. For tapered holes, a short length of the hole must be straight to accommodate the upper punch. Side holes or holes not parallel to the pressing direction generally require a secondary operation.

Lightening holes are used to reduce pressing area in large parts, lessening required pressing forces and final part weight. These advantages usually offset the additional tool cost, especially if round holes are used.

Length-to-Height Ratio

The length of a part should be proportional to its diameter. Excessively long parts are usually less dense in the center than at the ends. Usually, a length-to-height ratio of 3:1 or 4:1 is the practical limit, although ratios of 7:1 or 8:1 have been obtained where density distribution and strength were not critical.

It may be possible to redesign or modify a part to compensate for a variance in cross section. For example, a boss on one face, sufficiently high to make a significant difference in cross section, might be compensated for by a corresponding recess in the opposite face.

Multiple stepped parts can be molded, but the difference must be at least 0.032 in. to allow for sufficient tool member strength. It may be necessary to final machine some stepped sections.

Additional levels can be obtained without individual tool members if density variations can be tolerated. A hub is limited by the thickness of the collar and the projection of the collar from the main body of the part. To permit withdrawal of the upper punch, the slope of the edge of the hub should be 0.008 in./in. If the diameter of the part is less than 3/32 in., the height of the hub cannot exceed the height of the collar. If the diameter is greater than 3/32 in., the height of the hub can be 1½ times as high as the collar. Generally, height should be limited to 15% or less of the over-all part thickness in other bosses.

Splines

Cylindrical and symmetrical splines are preferred. Avoid splines that are narrow in relation to depth; dies for such splines are difficult to construct and costly to maintain. The width should not be less than half the length of the fin, and should never be less than 1/16 in. Radii at tips or bases of fins should not be less than 1/64 in.

Knurls and Serrations

Diamond knurls or horizontal serrations cannot be produced by pressing. Vertical serrations can be molded. To minimize tool costs, radii at the edges and bases of the serrations should be at least 0.008 in.

Chamfers or Bevels

Chamfers with maximum angles of 45° ending in shoulders are frequently used to break edges. Chamfers or bevels of less than 30° from the vertical cannot be molded, because the use of such angles will require construction of punches with feather edges. High-density work requires a

60° angle to minimize side loading of the punch. Any chamfer should have at least 0.010 in. at the tool leading edge. Assembly lead chamfers may sometimes be produced in the die or by a re-pressing operation.

Corner Reliefs

A relief can be formed at the junction of a flange and the body of the part if it runs in the pressing direction. Reliefs at right angles to the pressing direction are reentrants that cannot be pressed but must be machined after the part has been sintered.

Shapes That Cannot Be Molded

Although new techniques in tooling and side-motion compacting make it possible to produce certain contours with special equipment, the following details generally should be designed out of applications in order to minimize costs:

- Reentrant grooves, recesses and undercuts
- Any part having a taper in a direction that would form a reentrant angle
- Holes at right angles to the direction of pressing
- External and internal threads.

All of these contours usually must be introduced by secondary operations after the part has been sintered.

Tolerances

The nature of the P/M process and the variables that can be introduced during pressing and sintering have considerable influences on the tolerances that can be held. Closer tolerances can be held on radial than on longitudinal dimensions; radial dimensions are controlled by the accuracy imparted to the dies and punches, whereas longitudinal dimensions depend on the motion of the press and the accuracy of filling the die. Typical tolerances for ferrous parts of uniform cross section are presented in Table 1.

Variable cross sections and irregular shapes usually require more liberal tolerances. Increased tolerance control is expensive; to minimize costs, the design should specify only what is actually required.

Table 1. Typical tolerances for ferrous P/M parts of uniform cross section

	Typical tolerance, in./in.			
	As-sintered parts		Coined or sized parts	
Characteristic	Practical	Possible	Practical	Possible
Length	±0.005	±0.003	±0.004	±0.003
Inside diameter	±0.002	±0.001	±0.001	±0.0005
Outside diameter	±0.002	±0.001	±0.001	±0.0005
Concentricity (TIR)	0.003	0.002	0.002	0.0015
Flatness on ends	±0.002	±0.001	±0.001	±0.001
Parallelism of ends	±0.0015	±0.001	±0.001	±0.001

Generally, tolerances on powder metal applications compare favorably with those associated with conventional machining processes. Tolerances on as-sintered parts are comparable to turning, milling and boring tolerances. Sized powder metal parts have tolerances similar to medium grinding or broaching tolerances on dimensions perpendicular to the compacting direction.

Surface Finish

Porosity in powder metal parts affects the accuracy of surface finish measurements. The surface of a powder metal part consists of a series of flats and pores, with surface porosity increasing as part density decreases. Conventional measuring equipment cannot distinguish between this porosity and actual surface imperfections. Low-density parts can appear to have rough surfaces due to porosity. Development of a stylus that bridges porosity and thus improves accuracy of readings is in progress.

Additional Processing of Powder Metal Parts

Whenever possible, P/M parts should be designed so as to minimize secondary operations. However, if an application demands increased tolerances or density for better properties, or shapes not practical for powder metal processing, additional operations may be required.

Coining or re-pressing of the sintered part will increase density-related properties and increase tolerance control. In most cases, an additional set of tools will be required and piece price will increase due to the labor-intensive nature of the operation. Subassembly is also feasible during coining. Multiple powder metal parts can be fastened together to fabricate a subassembly while improving tolerances and density. Pins or shafts from a screw machine operation can be joined to powder metal parts (see Fig. 4). A screw machine pin with a knurled section is placed in a sintered hole. During coining, the thickness of the powder metal part is reduced, deforming material into the pin serrations. The strength of this junction is better than that of one produced by an interference press fit.

Joining of powder metal parts is possible during sintering (see Fig. 5). Here, two powder metal parts are assembled in the green state and sintered as an assembly. The outer section is made of a blend of material that will shrink during sintering; the inner portion is made of a material that grows. In the green state, the parts join in a slip fit; after sintering, the combination of growth and shrinkage of the parts forms a mechanical bond. The end result is a double-flanged hubbed pulley that is ready for machining.

Powder metallurgy parts can be staked satisfactorily (see Fig. 6). In this application, two inexpensive aluminum stampings have been assembled to an aluminum powder metal pulley, forming a double-flanged pulley that could not be produced entirely from powder metal. Multiple parts can also be staked during a coining operation to form assemblies (see Fig. 7). When staking or forming is required, the manufacturer must be made aware of this requirement in order to ensure correct properties.

Shapes or characteristics that cannot be feasibly produced by P/M techniques can be obtained by machining. Machining characteristics of powder metal are similar to those of cast iron and other cast materials. Best results are obtained by use of carbide tooling at high speeds and light feeds. Tapping poses little difficulty if plated or oversize taps are used to compensate for powder metal springback. Roll taps can be used successfully for making small holes in ductile materials that tend to snap standard cutting taps. Grinding is not recommended for bearing surfaces having self-lubricating properties, because the surface pores might become closed. If porosity is not critical, good grinding results are produced with hard, fine-grain wheels, operating at conventional speeds and light feeds. In using

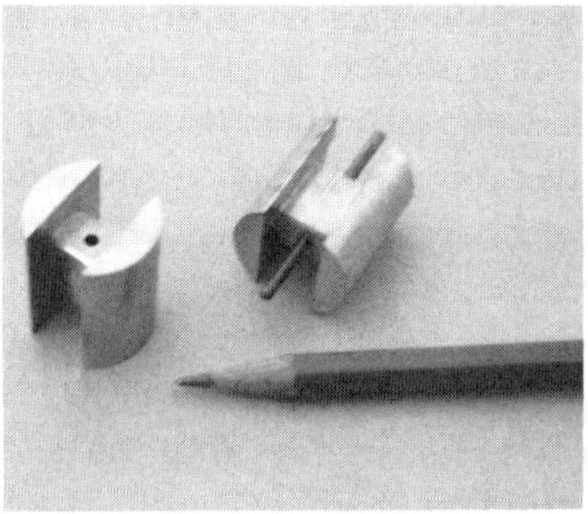

Fig. 4. Assembly consisting of a screw machine pin and a P/M part joined during coining.

Fig. 6. Assembly consisting of two aluminum stampings and an aluminum P/M pulley.

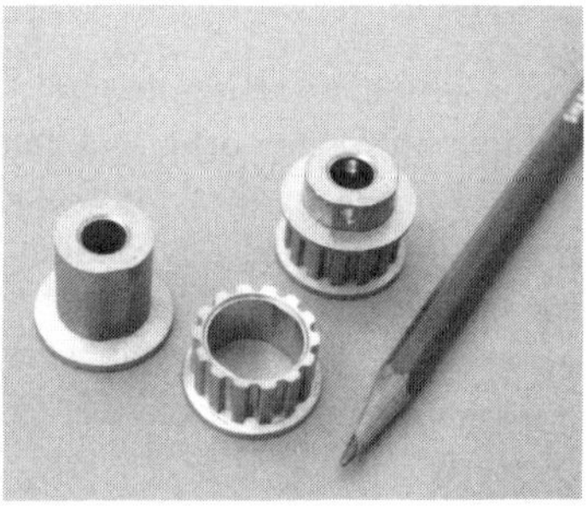

Fig. 5. Assembly consisting of two P/M parts joined during sintering.

Fig. 7. Assembly produced from multiple parts by staking during a coining operation.

coolants, care must be taken to avoid retention of the coolant, which may be detrimental to subsequent operations.

Plating or coating of powder metal parts can improve corrosion resistance and appearance. High-density parts are easily plated by standard procedures. In plating of lower-density parts, special procedures must be observed to keep plating salts from being entrapped in the pores of the component. Resin impregnation of the parts before plating is usually the most economical method of pore closure. Mechanical components are being plated with copper, nickel, chromium, cadmium, zinc and other metals.

Steam treating can increase general hardness, compressive strength, wear resistance and corrosion resistance of iron powder parts. Parts are processed by heating to about 1100°F and subjection to superheated steam under pressure. This forms a layer of magnetic iron oxide on all surfaces, including interconnected porosity. Other oxidizing and phosphating treatments may also be used.

High-density parts may be heat treated by conventional methods. Case control on powder metal parts is difficult except with high-density parts or alloys with low levels of interconnected porosity. Liquid carburizing of low-density components should be avoided because of possible entrapment of salt within the pores.

Summary

Powder metallurgy can be an effective method of reducing costs while maintaining component requirements. An effort has been made to explain the basics of design and to develop an interest in potential applications, with an emphasis on close cooperation with qualified parts producers to obtain optimum benefits. Powder metallurgy is not the answer to all problems, but can be of great assistance when used properly.

Chapter 4

Ferrous Powder Metallurgy

Part I: Pressed and Sintered Parts and Their Applications

Peter W. Lee
Senior Research Specialist
The Timken Company

Ever since the first commercial applications of the ferrous sintered oil pump gear in 1937/1938, the consumption of ferrous powder has increased steadily. Since 1960, powder consumption has increased at a compound rate of about 8% annually. Approximately 80% of ferrous powder is used in P/M part applications and, of the parts market, the automotive industry is a leading powder user. Other important areas are business machines, hand tools, lawn and garden equipment and appliances. The P/M process provides economical as well as technical benefits over conventional processing of wrought materials.

Some of the advantages of P/M processing are:

- **Reduction of Manufacturing Cost.** Generally, P/M processing provides higher material utilization, minimizes machining and reduces energy consumption. Consequently, over-all manufacturing cost is reduced.
- **Improved Performance.** Generally, the P/M process has the potential to improve performance through uniform

properties, fine grain structures and chemical homogeneity.

- **Design Flexibility**. P/M offers the designer material/property combinations that permit production of engineered structures for optimized mechanical and physical properties.
- **Production of Unique Materials.** P/M is the only viable manufacturing process for materials with very poor workability such as ceramics and carbides.

Manufacturing processes for pressed and sintered (P/S) P/M parts consist of powder production, cold compaction and sintering. The parts are used in applications requiring low mechanical property levels. Secondary processes, such as re-pressing and infiltration, can be used to improve the mechanical properties of the pressed and sintered P/M parts. For high-stress critical applications, the pressed and sintered P/M parts are hot forged into full-density parts.

Ferrous Metal Powders

Several commercial methods are available for production of ferrous powders. Among these, the most important processes are direct reduction of oxides and atomization of liquid metal.

Direct Reduction of Oxides

There are two commercially available processes for direct reduction of oxides: reduction with carbon (Hoeganaes Corp.) and reduction with hydrogen gas (Pyron Corp.).

In the Hoeganaes process, sponge iron is produced by direct reduction of magnetite (Fe_3O_4) ore[1]. Magnetite ore is dried in a rotary dryer and ground and cleaned with a magnetic separator. Cleaned ore is then charged into silicon carbide tubes with a crushed lime and coke mixture and heated at 2200°F for 24 to 36 hours in a tunnel kiln. Iron ore is reduced by carbon monoxide gas generated from the coke. The caked sponge iron is then removed from the container, ground and screened.

Impurities in the sponge iron powder are removed by a magnetic separator, and the cleaned powder is charged

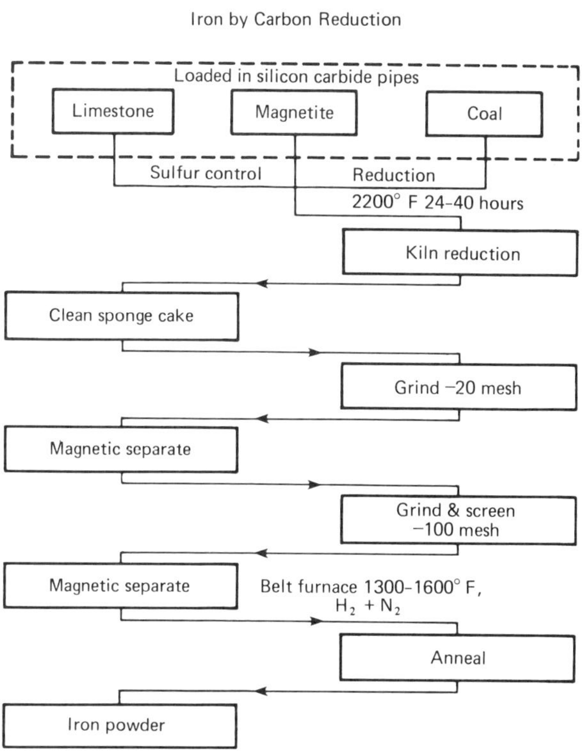

Fig. 1. Flow chart for production of Hoeganaes sponge iron powder[1]

into a strip belt furnace at 1700°F under a dissociated ammonia atmosphere in order to reduce the oxygen level of the powder. This process is shown schematically in Fig. 1.

In the Pyron process, iron powder is produced by reducing selected iron oxides under a hydrogen atmosphere. In this process, mill scale is ground, magnetically separated and screened. The mill scale is then preheated in air in order to convert all the oxides into hematite (Fe_2O_3), because hematite reduces more readily than magnetite (Fe_3O_4). The oxides are then reduced in a belt

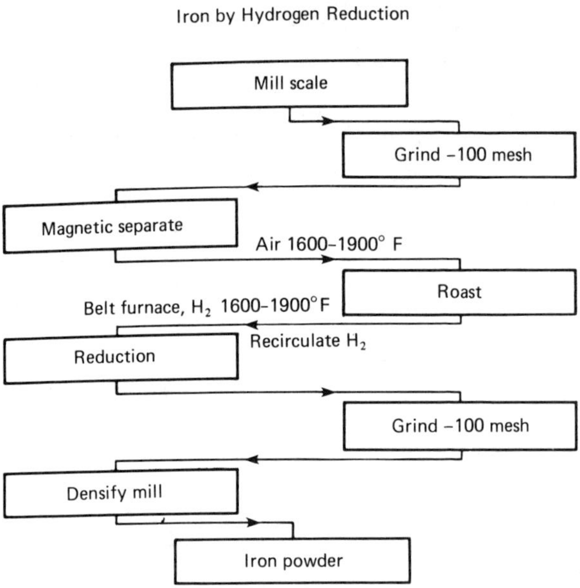

Fig. 2. Flow chart for production of Pyron iron powder

furnace under a hydrogen atmosphere at approximately 1800°F. The reduced iron is then ground, screened and blended to specifications. The Pyron process is shown schematically in Fig. 2.

The carbon-reduced Hoeganaes sponge iron particles have highly convoluted surfaces and contain internal porosity. The hydrogen-reduced Pyron iron powder particles are of irregular shape and contain high internal porosity. The internal pores of hydrogen-reduced iron powder particles are considerably finer than those of carbon-reduced sponge iron powder particles. Scanning electron micrographs of carbon-reduced Hoeganaes sponge iron powder and hydrogen-reduced Pyron iron powder are presented in Fig. 3.

Typical properties of Hoeganaes sponge iron powder and Pyron iron powder are given in Table 1. Generally, reduced iron powders are used for low- or medium-stress applications. Because of their irregular particle shape, reduced iron powders are readily die compacted.

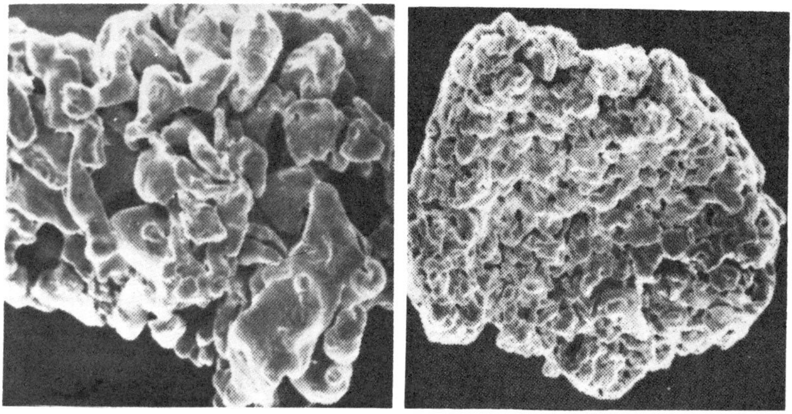

(a) (b)

Fig. 3. Scanning electron micrographs of particles of (a) Hoeganaes sponge iron powder and (b) Pyron iron powder[1]

Table 1. Typical compositions and properties of Hoeganaes (sponge) iron powder and Pyron iron powder

Element or property	Hoeganaes (sponge) iron powder	Pyron iron powder
Composition, %:		
Fe	98.2	97-98
C	0.02	0.01-0.05
S	0.01	0.005
P	0.01	0.012
H_2 loss	0.40	0.2-0.5
Apparent density, g/cm^3	2.50	2.3-2.5
Flow rate, s	30	27-35
Sieve analysis, %:		
+80 mesh (180 μm)	1	Trace
+100 mesh (150 μm)	4	2 max
+150 mesh (106 μm)	20	10-15
+200 mesh (75 μm)	27	15-25
+325 mesh (45 μm)	24	25-40
−325 mesh (45 μm)	24	28-45
Green density at 30 tsi, g/cm^3	6.4	6.3

Atomization of Liquid Metal

Another major commercial process for production of powder is atomization of liquid metal by either gas or water.

Gas Atomization. In the gas atomization process, a liquid metal stream is impinged upon by a subsonic or supersonic jet of inert gas. Atomization is produced by the kinetic energy of the atomizing fluid, typically nitrogen or argon for ferrous powders.

The particle size of the gas-atomized metal powder depends on several interrelated parameters such as jet pressure, jet distance, nozzle diameter, nozzle shape, gas velocity, and temperature of the liquid metal. Several empirical relations have been proposed to correlate the mean particle size of gas-atomized powder to processing parameters. Increasing jet distance and decreasing jet velocity tend to produce coarser particles. In gas atomization, the included angle of the gas stream with the melt stream is very important. Too large an angle produces lateral spray of the atomized particles and a high back pressure against the metal stream. Too small an angle of impingement usually means a long gas path, leading to gas turbulence and poor atomization efficiency.

Generally, gas atomization produces smooth and spherical powders, as shown in Fig. 4. It is employed for highly alloyed materials which contain significant amounts of readily oxidized alloy elements. Because of the spherical shape of gas-atomized powder particles, they can only be consolidated by applying pressure and heat simultaneously. Hot isostatic pressing, hot extrusion or containerized hot pressing such as ASEA's STAMP process is used for consolidation of gas-atomized powder.

Generally, inert-gas atomization and consolidation are expensive processes, and current applications are primarily limited to the production of tool steel, superalloy, titanium and other reactive metal powder parts. Detailed reviews of gas atomization have been published by Lubanska(2), Klar and Shafer(3), Dixon(4) and Rao(5).

Water Atomization. In water atomization, a high-pressure water stream is forced through a nozzle to form a

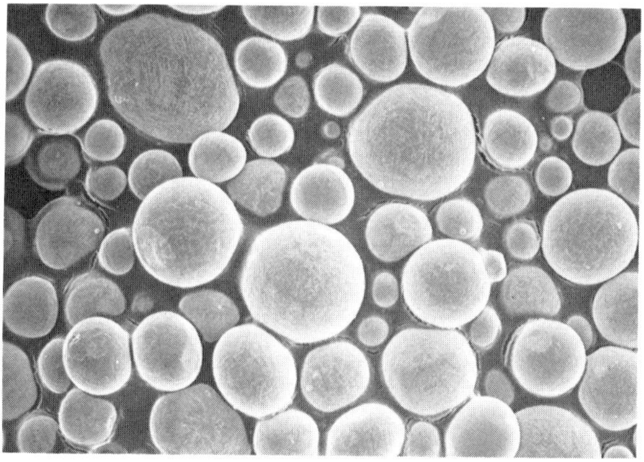

Fig. 4. Scanning electron micrograph of gas-atomized powder

dispersion of droplets which impact the metal stream. Generally, water-atomized powders have highly irregular particle shapes and high oxygen levels. As in gas atomization, several empirical relations between mean powder particle size and atomizing process conditions have been developed. The mean particle diameter decreases with increasing water droplet velocity and increases with the angle between metal stream and water jet. Some of the important process parameters in water atomization are:

- Thickness of water jet or stream
- Angle of water impingement on the metal stream
- Distance of water travel
- Water pressure
- Metal temperature
- Metal stream diameter
- Metal viscosity.

Grandzol[6] made an extensive study of water atomization and concluded that the formation of metal droplets during water atomization was the result of impact, not shear, of the water droplets on the surface of the metal stream.

A scanning electron micrograph of water-atomized powder is shown in Fig. 5. Because of the highly irregular particle shape, water-atomized powder can be compacted

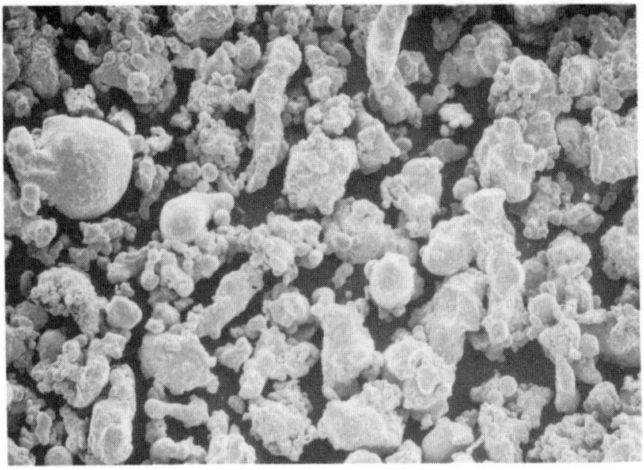

Fig. 5. Scanning electron micrograph of water-atomized powder

at room temperature by either die pressing or cold isostatic pressing. Prealloyed powders are produced by atomization processes, and, consequently, atomized powders are used for high-stress and high-performance applications.

Compaction and Sintering Technology

The Compaction Process

After blending, the compaction of powder into green parts is the next major step of the P/M process. The Metal Powder Industry Federation (MPIF) has classified powder metallurgy parts into the following four groups, depending on their design complexity. Class I parts are one-level parts which have a thickness of less than 0.25 in. and which can be pressed with a force from one direction. Class II parts are one-level parts of any thickness that must be pressed with forces from two opposing directions. Class III parts are two-level parts of any thickness that must be pressed with forces from two opposing directions. Class IV parts are multilevel parts of any thickness that must be pressed with forces from two opposing directions.

The reduced iron powders and water-atomized powders are compacted into preforms by either die compaction or cold isostatic pressing. The irregularly shaped powder particles are mechanically interlocked during compaction and provide substantial green strength in the compact.

Die compaction is probably the most widely used technique for P/M part production because of its speed and ability to produce fairly complex parts. However, highly complex P/M parts are produced by cold isostatic pressing. Some of the terminology used in the die compaction process is as follows:

Pressing area is the largest cross-sectional area perpendicular to the pressing direction.

Ejection surface area is the surface area of the part parallel with the pressing direction.

Compression ratio is the ratio of the apparent density of the powder charge to the green density of the compact.

Powder fill depth is the depth of die cavity filled with loose powder to produce compacts of the required density level.

Apparent density is the density of free-flowing metal powder.

Green density is the density of the pressed part.

Final density is the density of the final part.

Neutral axis is the low-density plane in the green compact that is perpendicular to the direction of pressing.

Solid lubricant and graphite powders are often blended with ferrous powder. A lubricant reduces the required ejection force, and graphite provides carbon for reduction of oxides and the final carbon level of the sintered part. Since the solid lubricant tends to reduce green strength of the compacts and to weaken the final sintered parts, a minimum amount of lubricant is recommended. Die wall lubrication can be used instead of a lubricant blended into the powder mass to minimize degradation of properties by the presence of residual lubricant.

During die compaction, friction develops from the relative motion between die sidewalls and powder particles, between powder particles within the powder mass, and between powder particles and punch faces. Because of this friction, the local pressure within the powder mass caused by the upper punch load is not uniform, which results in density variations in the compact. For thick P/M parts, forces should be applied from both top and bottom directions to control the neutral axis. The density variations for both single- and double-action compaction are shown schematically in Fig. 6. In addition, since metal powder does not flow under pressure, separate pressing forces must

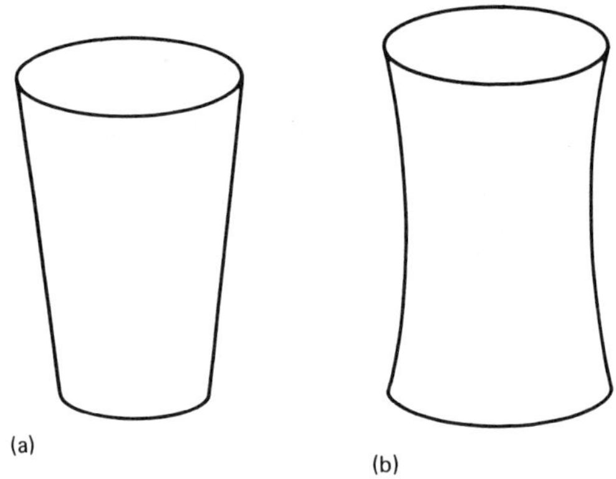

Fig. 6. Sintered green compacts made by (a) single-action and (b) double-action pressing, showing shape distortions that occurred during sintering as a result of density variations in the green compacts

be provided for each level in multilevel parts. Even for single-level parts, uniform filling of the die cavity is important for obtaining compacts with uniform density. Uniform density in the compact is necessary to obtain uniform density in the final part after subsequent sintering operations.

The life of the compaction tooling is influenced by the required tolerances of the part. Close tolerances tend to decrease tool life by decreasing the amount of tool wear that can be accommodated. In addition, closer tolerances may require secondary operations such as coining or sizing, which increase costs. Therefore, it is recommended that tolerances of P/M parts be as liberal as possible. Generally, the following tolerances can be achieved on parts in the as-pressed and as-sintered conditions[7]:

- Length (in direction of pressing):
  1 in. or less . . . ±0.002
  1 to 3 in. . . . . . ±0.005
  Over 3 in. . . . . . ±0.010
- Radial (at right angles to direction of pressing):
  ±0.001 to 0.002 in./in.
- Cored holes:
  ±0.0005 to ±0.001 in.
- Concentricity:
  Up to 1-in. bore . . . 0.002 in. TIR
  1 to 2 in. . . . . 0.003 in. TIR
  2 to 3 in. . . . . 0.004 in. TIR
  Over 3 in. . . . . 0.006 in. TIR
- Gear tolerances are usually based on AGMA classification numbers. In normal processing, powder metallurgy gears can be held to AGMA quality numbers 5 to 7. By utilizing special sizing techniques, gears of AGMA quality number 8 can be produced.
- Sizing after sintering will give closer tolerances but at added cost. For example, length tolerances in the direction of pressing can be held to ±0.001 in. Hole diameter tolerances can usually be held to ±0.0005 in.

Cold Isostatic Compaction. In isostatic pressing of P/M parts, equal pressure is applied in all directions on a metal powder compact. Generally, isostatic pressing of ferrous P/M parts has rather limited use because of its low production speed. However, isostatic pressing has advantages and should be considered as an alternative to die compaction.

Some of the advantages of isostatic pressing are as follows:

- Because uniform pressure is applied in all directions and no die wall friction is present, an isostatically pressed compact has a higher and more uniform density than a die pressed compact.
- Due to the absence of solid lubricant, an isostatically pressed compact has higher green strength and green density than a die pressed compact.

- Because of the hydrostatic stress presented in isostatic pressing, materials which have poor workability, such as brittle ceramic compounds and refractory metals, can be pressed.
- Isostatic pressing is preferred for parts with large ratios of height to cross-sectional area.
- Because of the absence of vertical ejection, parts with complex shapes which are impossible to make by die compaction can be made by isostatic pressing.

Isostatic pressing involves the following basic steps:

- The powder is filled into a flexible mold having a shape similar to the final part shape. The mold is usually made of rubber, urethane, polyvinyl chloride or other elastomers, and the initial size is determined by the compression ratio of the powder.
- Air in the mold is evacuated and the mold is sealed. De-airing is optional for most ferrous powders.
- Isostatic pressure is applied to the sealed mold via water, oil or a combination of both.
- The pressure is released and the compacted P/M part is removed from the mold.

Two types of isostatic pressing—namely, "wet bag" and "dry bag" processing—are available. Wet bag processing, or the free mold method, involves removal of the mold from the pressure vessel after pressing. Mold filling and stripping are performed outside of the pressure vessel. In dry bag processing, the mold is attached to the pressure vessel and only the pressed part is removed after pressing. Automated systems for high-volume production are based on the dry bag approach. Both wet bag and dry bag processing are illustrated schematically in Fig. 7.

The Sintering Process

Sintering is probably one of the most complex phenomena in P/M processing. The term "sintering" may have many different meanings to different people. For powder metallurgy applications, sintering may be generally defined as the process wherein powder particles develop metallurgical bonding and densify under the influence of heat.

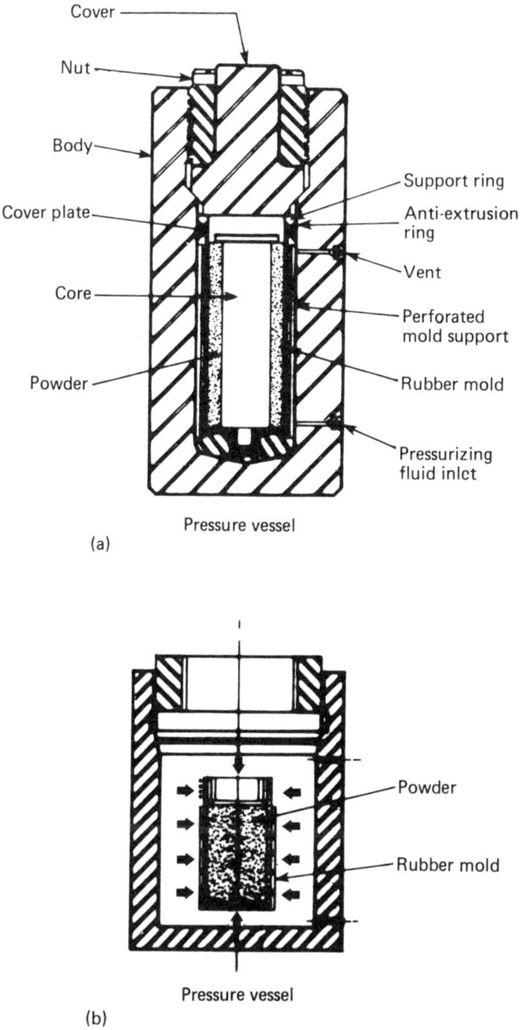

Fig. 7. Schematic illustrations of isostatic pressing techniques. (a) Dry bag method. (b) Wet bag method.

During sintering of ferrous P/M parts, in particular, the following reactions take place:

- Metallurgical bonds are developed from the mechanical interlocks between powder metal particles in the compact.
- Metal oxides in the powder compact are reduced by reaction with the carbon from the blended graphite powders.
- The desired final carbon level of the P/M compact is obtained by diffusion of carbon from graphite powder.
- Densification of P/M compacts can be achieved during sintering. The degree of densification depends on the sintering parameters and the alloy(s) involved. Generally, higher sintering temperatures and longer sintering times promote densification of sintered parts.

Based on the geometrical changes that occur during sintering, the sintering stages are considered to consist of (*a*) initial bonding between particles and neck growth, (*b*) pore channel closure and pore rounding, and (*c*) densification and pore coarsening[8]. The driving force for sintering is minimization of surface energy by reduction of surface area. Hence, the greater the amount of internal surface area in the green compact, the greater the driving force during sintering. A typical sintering cycle for ferrous powder consists of preheating or delubrication at about 1300°F and sintering at 2050°F for 30 minutes in a reducing atmosphere, followed by cooling under inert atmosphere. In recent years, however, high-temperature sintering has become increasingly popular because of the improvement in dynamic mechanical properties and reduction in oxygen level. In high-temperature sintering, green compacts are heated at temperatures of 2250°F or higher. Sintering at 2050°F is referred to as conventional sintering.

The commercially used sintering atmospheres are endothermic, exothermic, dissociated ammonia, pure hydrogen, and nitrogen-base atmospheres. Approximate compositions of these gases are given in Table 2. Among these atmospheres, endothermic gas is probably the most widely used, followed by dissociated ammonia. A pure hydrogen atmosphere has rather limited use (mainly for aerospace applications) due to its cost. Nitrogen-base atmospheres are fairly new for powder metallurgy applications; however, their application increases steadily because of low cost and extremely low dew point.

Endothermic gas is readily produced from natural gas at reasonable cost. However, control of carbon level is difficult because the carbon potential of endothermic gas varies with temperature and dew point, as shown in Fig. 8. The carbon potential of the endothermic gas with +30°F dew point varies from approximately 0.2% at 2050°F to 1.0% at about 1560°F.

Dew point is a measure of the dryness of the atmosphere gas. The lower the dew point, the smaller the water vapor content of the atmosphere. Generally, low dew point gases will increase the reduction potential of the atmosphere. The relationship between dew point and water vapor content is given in Fig. 9.

A typical sintering furnace consists of three zones: a preheat or delubricating zone, a high-heat or sintering zone, and a cooling zone. For conventional sintering at 2050°F, mesh belt conveyor furnaces are the most widely used. Green compacts are carried on a conveyor mesh belt made from nickel-chromium alloy wire. Mesh belt furnaces can be used at temperatures up to 2100°F only, due to the temperature limitation of the belt material. For high-temperature sintering applications, walking beam furnaces have been successfully used for continuous operation at temperatures up to 2400°F. Vacuum furnaces can also be used for high-temperature batch-type sintering of ferrous P/M parts. However, any lubricant blended into P/M compacts must be burned off in a separate controlled-atmosphere furnace before the compacts are sintered in a vacuum furnace.

Typical Properties of P/S Parts

One of the unique characteristics of pressed and sintered P/M parts is their residual porosity. It is well known that fractures of fully dense conventional material during plastic deformation are initiated at nonmetallic inclusions or other structural discontinuities. Edelson and Baldwin[9] demonstrated that tensile ductility decreases with increasing fraction of the second-phase particles or holes.

Table 2. Typical compositions of sintering atmospheres[7]

Atmosphere gases	Approximate air: natural gas ratio	Dew point, °F, upon entering furnace	Properties at elevated temperatures(a)	Composition, %: Carbon dioxide (CO_2)	Oxygen (O_2)	Carbon monoxide (CO)	Hydrogen (H_2)	Methane (CH_4)	Nitrogen (N_2)	Argon (A)	Helium (He)
Hydrogen											
By electrolysis of water (direct from cells)											
Unpurified–saturated	...	+70 to +90	R, r, D3	0.0	0.2	0.0	99.8	0.0	0.0	0.0	0.0
Purified	...	–80 or lower	R, r, *N*	0.0	0.0	0.0	100.0	0.0	0.0	0.0	0.0
From bottles											
Unpurified	...	–30 to +20	R, r, D1	0.0	0.0	0.0	99.95	0.0	0.0	0.0	0.0
Purified	...	–80 or lower	R, r, *N*	0.0	0.0	0.0	100.0	0.0	0.0	0.0	0.0
By catalytic conversion of hydrocarbons											
Unpurified–saturated	...	+70 to +90	R, r, D3	0.05	0.0	0.02	99.73	0.20	0.0	0.0	0.0
Dried	...	–100 or lower	R, r, *N*	0.002	0.0	0.02	99.97	0.0	0.0	0.0	0.0
From liquid hydrogen	...	–84 or lower	R, r, *N*	0.0	0.0	0.0	99.998	0.0	0.0	0.0	0.0
Nitrogen–bottled	...	–64 or lower	O, o, D1	0.0	0.0	0.0	0.0	0.0	99.996	0.0	0.0
Hydrogen-nitrogen mixtures											
Dissociated ammonia											
As reacted–dry	...	–60 to –40	R, r, *N*	0.0	0.0	0.0	75.0	0.0	25.0	0.0	0.0
Moisture added–saturated	...	+70 to +100	R, r, D3	0.0	0.0	0.0	75.0	0.0	25.0	0.0	0.0
Burned dissociated ammonia											
Rich–saturated	...	+70 to +90	R, r, D3	0.0	0.0	0.0	24.0	0.0	76.0	0.0	0.0
Lean–saturated	...	+70 to +90	O, n, D3	0.0	0.0	0.0	1.0	0.0	99.0	0.0	0.0
Direct catalytic conversion of ammonia and air											
Rich											
As reacted–saturated	...	+70 to +90	R, r, D3	0.0	0.0	0.0	25.0	0.0	75.0	0.0	0.0
Cooled to 40°F		+40	R, r, D2	0.0	0.0	0.0	25.0	0.0	75.0	0.0	0.0

[illegible]	...	+70 to +90	O, n, D3	0.0	0.0	0.0	1.0	0.0	99.0	0.0	0.0
Cooled to 40°F	...	+40	O, n, D2	0.0	0.0	0.0	1.0	0.0	99.0	0.0	0.0
Dried	...	−80 or lower	N, n, *N*	0.0	0.0	0.0	1.0	0.0	99.0	0.0	0.0
Reformed hydrocarbon gases											
Exothermic gas											
Rich											
As reacted–saturated	6:1	+70 to +90	R, r, D3	5.0	0.0	10.0	14.0	1.0	70.0	0.0	0.0
Refrigerated	6:1	+40 to +60	R, r, D2	5.0	0.0	10.0	14.0	1.0	70.0	0.0	0.0
Medium rich–saturated	6.75:1	+70 to +90	O, r, D3	6.3	0.0	8.8	11.5	0.3	73.1	0.0	0.0
Lean–saturated	10.25:1	+70 to +90	O, n, D3	11.5	0.0	0.7	0.7	0.0	87.1	0.0	0.0
Purified exothermic gas											
Rich	6:1	−40 or lower	R, r, C1	0.0	0.0	10.8	15.0	1.0	73.2	0.0	0.0
Medium rich											
As reacted	6.75:1	−40 or lower	R, r, C1	0.0	0.0	9.0	12.0	0.2	78.8	0.0	0.0
Methane added	6.75:1	−40 or lower	R, r, C3	0.0	0.0	8.5	11.4	5.2	74.9	0.0	0.0
Lean	10.25:1	−40 or lower	N, n, *N*	0.0	0.0	0.7	0.7	0.0	98.6	0.0	0.0
Endothermic gas											
Rich–dry	2.4:1	−10 to +10	R, r, C3	0.0	0.0	20.0	38.0	0.5	41.5	0.0	0.0
Rich–fairly dry											
As reacted	2.6:1	+20 to +30	R, r, C2	0.0	0.0	19.0	37.0	0.0	43.7	0.0	0.0
Methane added	2.6:1	+20 to +30	R, r, C3	0.0	0.0	18.1	35.2	5.3	41.4	0.0	0.0
Medium rich–saturated	3.5:1	+70 to +90	R, r, D1	1.2	0.0	16.5	27.6	0.0	54.7	0.0	0.0
Lean–saturated	4.5:1	+70 to +90	R, r, D3	3.0	0.0	13.8	21.5	0.0	61.7	0.0	0.0
Argon–bottled	...	−73 or lower	N, n, *N*	0.0	0.0	0.0	0.0	0.0	0.0	99.996	0.0
Helium–bottled	...	−73 or lower	N, n, *N*	0.0	0.0	0.0	0.0	0.0	0.0	0.0	99.995
Vacuum (below 150 μm)	...	...	R, r, *N*	0.0	0.0	0.0	0.0	0.0	0.0	0.0	0.0
Air											
Normal	...	+50 or lower	O, o, D3	0.0	21.0	0.0	0.0	0.0	78.1	0.9	0.0
Wet	...	Above +50	O, o, D3	0.0	21.0	0.0	0.0	0.0	78.1	0.9	0.0

(a) Properties at elevated temperatures with iron or iron oxides: O–oxidizing, N–neutral, R–reducing; with copper or copper oxides: o–oxidizing, n–neutral, r–reducing; with iron or iron-carbon: C1–mildly carburizing, C2–average carburizing, C3–strongly carburizing, *N*–neutral, D1–mildly decarburizing, D2–average decarburizing, D3–strongly decarburizing.

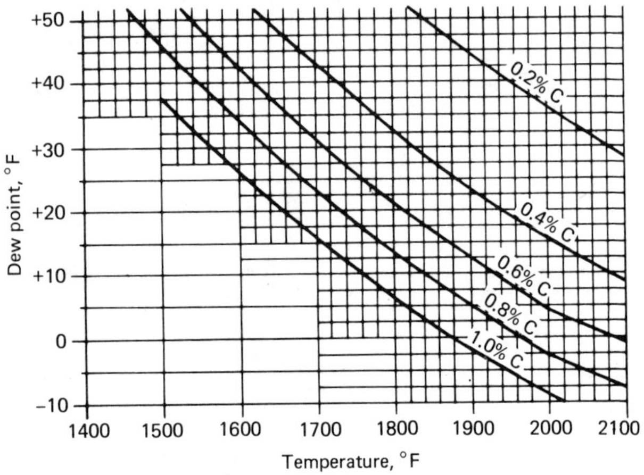

Fig. 8. Carbon potential of endothermic gas as a function of dew point and temperature

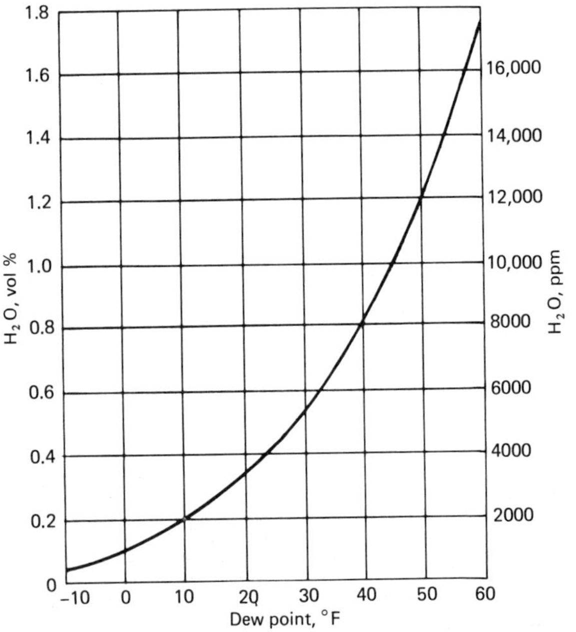

Fig. 9. Relationship between dew point and water vapor content

In porous P/M materials, both static and dynamic mechanical properties are strongly dependent on the degree of porosity. Essentially, the porous material already has cracks throughout its entire volume. The pores act as stress raisers under the presence of an external stress, which enhances crack propagation between pores to reduce the effective cross-sectional area of solid material. Generally, static properties such as tensile strength, yield strength, ductility and elastic modulus increase with decreasing porosity. Variations in the mechanical properties of sintered ferrous compacts made from six different powders are given as a function of density in Fig. 10 and 11[10,11]. Because of the presence of porosity, the apparent hardness of porous P/M parts is lower than the actual particle hardness. The relationship between apparent hardness and particle hardness for the various density levels is given in Fig. 12. The MPIF specifications for ferrous P/M structural materials and their density designations and ranges are given in Tables 3 and 4, respectively[12]. The typical mechanical properties obtainable at various density levels in different iron-base P/M materials are given in Table 5[13].

Typical Applications of P/S Parts

Presently, the major tonnage of ferrous powder is being used for P/M part applications. Among the part markets, automotive and off-road and construction equipment account for about 70% of the total volume.

Gear pumps constitute one of the most common types of hydraulic units. Depending on application stress levels, different materials and density levels are used. For example, iron-carbon or iron-carbon-copper materials with densities of 6.0 to 6.8 g/cm^3 are used in gear pumps for low-pressure applications such as engine lubrication and automatic transmissions. For higher stress applications up to 3000 psi, alloy steels at the minimum density of 7.1 g/cm^3 are used. P/M material requirements for gear pump applications at various stress levels are given in Table 6[14].

Applications of P/M parts in outdoor garden equipment have been increasing, primarily due to economics as well as flexibility of design capabilities. Some of the P/M parts used are reduction gears at a density of 7.3 to 7.5 g/cm^3, brake drums at a density of 6.4 to 6.6 g/cm^3, speed-change

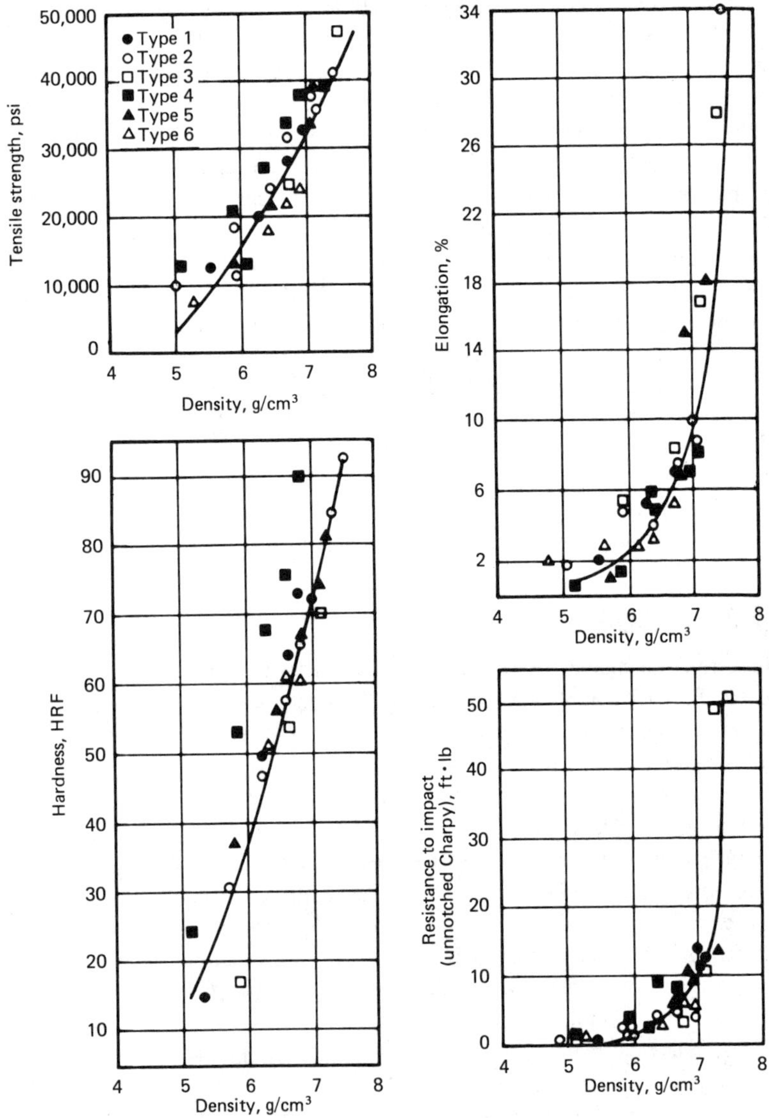

Fig. 10. Mechanical properties of iron compacts as a function of sintered density[(10)]

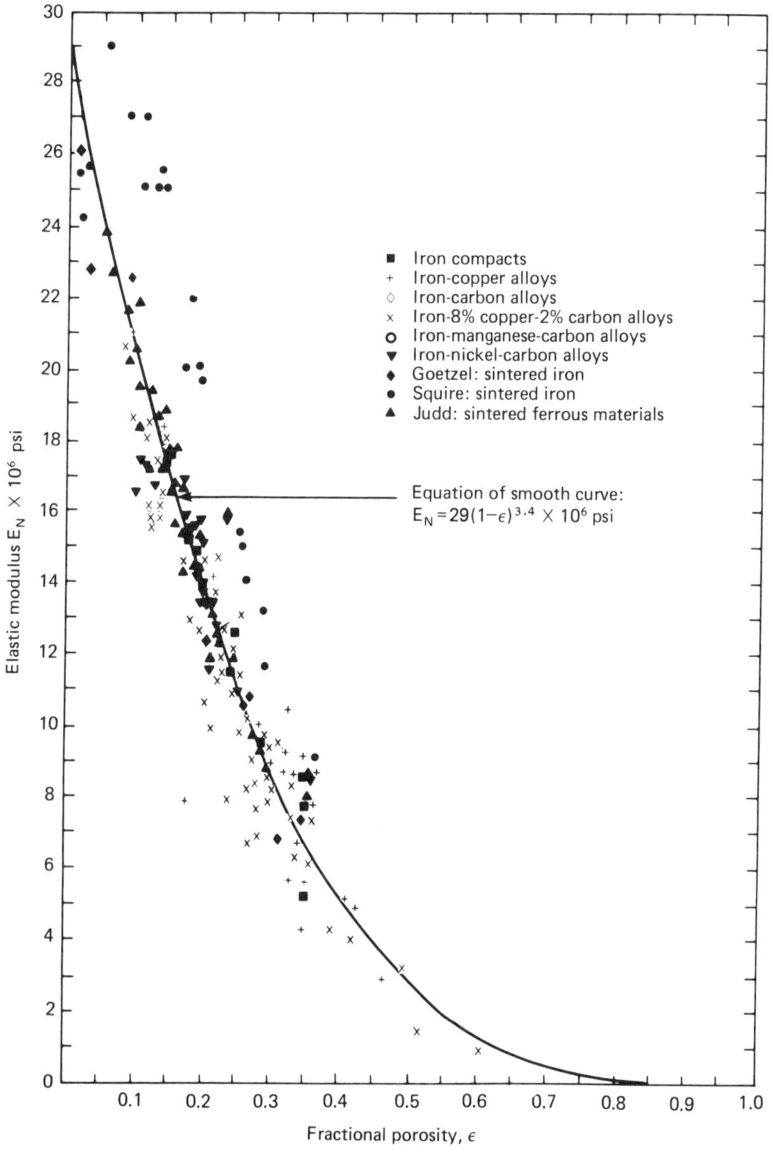

Fig. 11. Elastic modulus of ferrous sintered materials as a function of density[11]

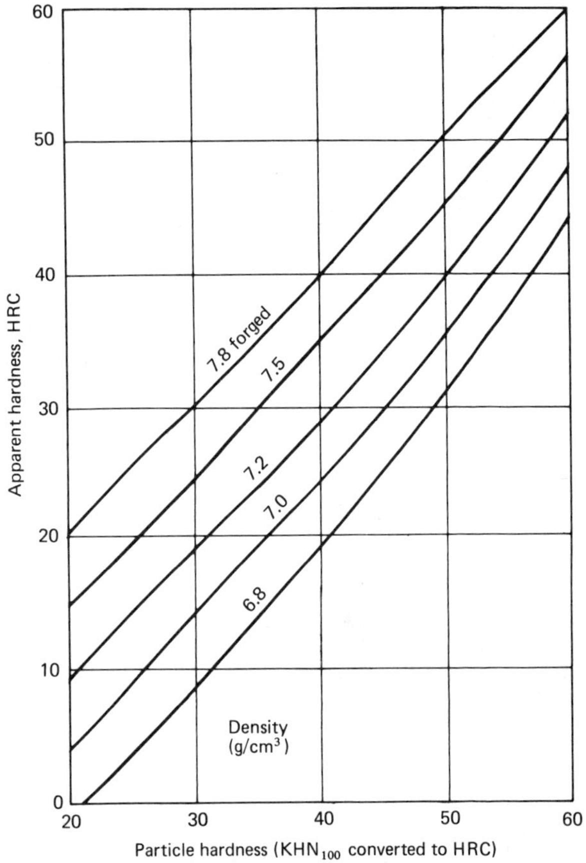

Fig. 12. Relationships between apparent hardness and particle hardness at various density levels. (Source: A. O. Smith Technical Data Sheet #1030, 1973)

gears at a density of 7.0 to 7.2 g/cm^3, and sprocket assemblies at a density of 6.8 to 7.0 g/cm^3[15].

Numerous P/M parts are being used in automotive applications. Pressed and sintered P/M automotive parts now in use include clutch pressure plates, crankshaft pulleys, and components of a variable displacement pump[16]. The

Table 3. Compositions of ferrous P/M materials(12)

Material	MPIF designation	MPIF composition ranges (a), % C	Ni	Cu
P/M iron	F-0000	0.3 max	...	...
P/M steel	F-0005	0.3-0.6	...	...
P/M steel	F-0008	0.6-1.0	...	...
P/M copper-iron	FC-0200	0.3 max	...	1.5-3.9
P/M copper steel	FC-0205	0.3-0.6	...	1.5-3.9
P/M copper steel	FC-0208	0.6-1.0	...	1.5-3.9
P/M copper steel	FC-0505	0.3-0.6	...	4.0-6.0
P/M copper steel	FC-0508	0.6-1.0	...	4.0-6.0
P/M copper steel	FC-0808	0.6-1.0	...	6.0-11.0
P/M iron-copper	FC-1000	0.3 max	...	9.5-10.5
P/M iron-nickel	FN-0200	0.3 max	1.0-3.0	2.5 max
P/M nickel steel	FN-0205	0.3-0.6	1.0-3.0	2.5 max
P/M nickel steel	FN-0208	0.6-0.9	1.0-3.0	2.5 max
P/M iron-nickel	FN-0400	0.3 max	3.0-5.5	2.0 max
P/M nickel steel	FN-0405	0.3-0.6	3.0-5.5	2.0 max
P/M nickel steel	FN-0408	0.6-0.9	3.0-5.5	2.0 max
P/M iron-nickel	FN-0700	0.3 max	6.0-8.0	2.0 max
P/M nickel steel	FN-0705	0.3-0.6	6.0-8.0	2.0 max
P/M nickel steel	FN-0708	0.6-0.9	6.0-8.0	2.0 max

(a) Remainder Fe.

Table 4. Density designations and ranges of ferrous P/M materials(12)

MPIF suffix	Density range, g/cm^3
N	<6.0
P	6.0 to <6.4
R	6.4 to <6.8
S	6.8 to <7.2
T	7.2 to <7.6
U	7.6 to <8.0
W	8.0 to <8.4

clutch pressure plates shown in Fig. 13 are made from iron-carbon-copper powders with densities of 6.7 to 6.8 g/cm^3. This is a P/M application of long standing. Figure 14 shows a crankshaft pulley which was used for driving the timing belt in a high-performance version of an American automotive engine. These pulleys are made from iron-

Table 5. Typical mechanical properties of ferrous P/M materials(13)

Nominal composition, %	MPIF designation	Density, g/cm^3	Condition(a)	Tensile strength, ksi (MPa)	Yield strength, ksi (MPa)	Elongation(b), %	Modulus of elasticity, 10^6 psi (10^4 MPa)	Apparent hardness	Impact strength (smooth Charpy), ft·lb (J)
Iron									
99.7-100Fe	F-0000-N	5.6-6.0	A.S.	16 (110)	11 (76)	2.0	10.5 (7.23)	10R$_b$	13.0 (4.07)
99.7-100Fe	F-0000-P	6.0-6.4	A.S.	19 (131)	14 (96)	2.5	13.0 (8.96)	70R$_b$	4.5 (6.1)
99.7-100Fe	F-0000-R	6.4-6.8	A.S.	24 (165)	16 (110)	5.0	16.0 (11.0)	80R$_b$	9.5 (12.9)
99.7-100Fe	F-0000-S	6.8-7.2	A.S.	30 (207)	21.5 (148)	9.0	19.0 (13.1)	15R$_b$	15.0 (20.3)
99.7-100Fe	F-0000-T	7.2-7.6	A.S.	40 (276)	26 (179)	15.0	25.0 (17.2)	30R$_b$	23.0 (31.2)
Steel									
97.4/99.7Fe-0.3/0.6C	F-0005-N	5.6-6.0	A.S.	18 (124)	14.5 (100)	1.0	10.5 (7.23)	5R$_b$	2.5 (3.4)
	F-0005-P	6.0-6.4	A.S.	25 (172)	20 (138)	1.5	13.0 (8.96)	20R$_b$	3.5 (4.7)
	F-0005-R	6.4-6.8	A.S.	32 (221)	23 (159)	2.5	16.0 (11.0)	45R$_b$	5.0 (6.8)
	F-0005-R	6.4-6.8	H.T.	60 (414)	57 (393)	<0.5	...	100R$_b$	...
	F-0005-S	6.8-7.2	A.S.	43 (296)	28 (193)	3.5	19.0 (13.1)	60R$_b$	9.0 (12.2)
	F-0005-S	6.8-7.2	H.T.	80 (552)	75 (517)	<0.5	...	25R$_c$	...
97.0/99.1Fe-0.6/1.0C	F-0008-N	5.6-6.0	A.S.	29 (200)	25 (172)	0.5	10.5 (7.23)	35R$_b$	2.0 (2.7)
	F-0008-N	5.6-6.0	H.T.	42 (290)	...	<0.5	...	90R$_b$	...
97.0/99.4Fe-0.6/1.0C	F-0008-P	6.0-6.4	A.S.	35 (241)	30 (207)	1.0	13.0 (8.96)	50R$_b$	3.0 (4.07)
	F-0008-P	6.0-6.4	H.T.	58 (400)	...	<0.5	...	100R$_b$	...
	F-0008-R	6.4-6.8	A.S.	42 (290)	36 (248)	1.5	16.0 (11.0)	65R$_b$	3.5 (4.7)
	F-0008-R	6.4-6.8	H.T.	74 (510)	...	<0.5	...	25R$_c$	...
	F-0008-S	6.8-7.2	A.S.	57 (393)	40 (276)	2.5	19.0 (13.1)	75R$_b$	7.0 (9.5)

Fe-2.7Cu-0.15C	FC-0200-P	6.0-6.4	A.S.	23 (159)	17 (117)	2.5	13.0 (8.96)	$80R_h$	5.5 (7.5)
	FC-0200-R	6.4-6.8	A.S.	30 (207)	21 (145)	4.0	16.0 (11.0)	$15R_b$	7.0 (9.5)
	FC-0200-S	6.8-7.2	A.S.	37 (255)	23 (159)	7.0	19.0 (13.1)	$30R_b$	17 (23.0)
Copper steel									
Fe-2.7Cu-0.45C	FC-0205-P	6.0-6.4	A.S.	40 (276)	34 (234)	1.0	13.0 (8.96)	$45R_b$	3.5 (4.7)
	FC-0205-R	6.4-6.8	A.S.	50 (345)	38 (262)	1.5	16.0 (11.0)	$70R_b$	5.5 (7.5)
	FC-0205-R	6.4-6.8	H.T.	85 (586)	81 (558)	<0.5	...	$30R_c$	...
	FC-0205-S	6.8-7.2	A.S.	62 (427)	45 (310)	3.0	19.0 (13.1)	$80R_b$	9.5 (12.9)
	FC-0205-S	6.8-7.2	H.T.	100 (689)	95 (655)	<0.5	...	$35R_c$	...
Fe-2.7Cu-0.8C	FC-0208-N	5.6-6.0	A.S.	33 (227)	30 (207)	<0.5	10.5 (7.23)	$45R_b$	2.5 (3.4)
	FC-0208-N	5.6-6.0	H.T.	43 (296)	...	<0.5	...	$95R_b$	...
	FC-0208-P	6.0-6.4	A.S.	45 (310)	41 (283)	<0.5	13.0 (8.96)	$60R_b$	3.0 (4.07)
	FC-0208-P	6.0-6.4	H.T.	55 (379)	...	<0.5	...	$25R_c$	...
	FC-0208-R	6.4-6.8	A.S.	60 (414)	48 (331)	1.0	16.0 (11.0)	$70R_b$	5.0 (6.8)
	FC-0208-R	6.4-6.8	H.T.	80 (552)	...	<0.5	...	$35R_c$	...
	FC-0208-S	6.8-7.2	A.S.	80 (552)	57 (393)	1.5	19.0 (13.1)	$80R_b$	8.0 (10.8)
	FC-0208-S	6.8-7.2	H.T.	100 (689)	95 (655)	<0.5	...	$40R_c$	...
Fe-5.0Cu-0.45C	FC-0505-N	5.6-6.0	A.S.	35 (241)	30 (207)	0.5	10.5 (7.23)	$50R_b$	3.0 (4.07)
	FC-0505-N	5.6-6.0	H.T.	...	...	...	...	$90R_b$	...
	FC-0505-P	6.0-6.4	A.S.	50 (345)	42 (290)	1.0	13.0 (8.96)	$60R_b$	4.5 (6.1)
	FC-0505-P	6.0-6.4	H.T.	...	...	...	...	$95R_b$	...
	FC-0505-R	6.4-6.8	A.S.	66 (455)	55 (379)	1.5	16.0 (11.0)	$75R_b$	5.0 (6.8)
	FC-0505-R	6.4-6.8	H.T.	...	...	...	16.0 (11.0)	$25R_c$	...
Fe-0.5Cu-0.8C	FC-0508-N	5.6-6.0	A.S.	48 (331)	43 (296)	<0.5	10.5 (7.23)	$60R_b$	3.0 (4.07)
	FC-0508-N	5.6-6.0	H.T.	...	...	...	...	$95R_b$	...
	FC-0508-P	5.6-6.0	A.S.	62 (427)	57 (393)	<1.0	13.0 (8.96)	$65R_b$	3.5 (4.7)
	FC-0508-P	5.6-6.0	H.T.	...	...	...	...	$30R_c$	...
	FC-0508-R	6.4-6.8	A.S.	75 (517)	70 (483)	1.0	16.0 (11.0)	$85R_b$	4.5 (6.1)
	FC-0508-R	6.4-6.8	H.T.	...	...	...	...	$35R_c$	...

continued

Table 5, continued

Nominal composition, %	MPIF designation	Density, g/cm³	Condition(a)	Tensile strength, ksi (MPa)	Yield strength, ksi (MPa)	Elongation(b), %	Modulus of elasticity, 10^6 psi (10^4 MPa)	Apparent hardness	Impact strength (smooth Charpy), ft·lb (J)
Copper steel, *contd.*									
Fe-8.5Cu-0.8C	FC-0808-N	5.6-6.0	A.S.	36 (248)	. . .	<0.5	. . .	55R_b	. . .
Fe-10Cu-0.15C	FC-1000-N	5.6-6.0	A.S.	30 (207)	. . .	0.5	. . .	70R_f	. . .
Iron-nickel									
Fe-2.0Ni-1.25Cu-0.15C	FN-0200-R	6.4-6.8	A.S.	28 (193)	18 (124)	4.0	17.0 (11.7)	38R_b	14.0 (18.9)
	FN-0200-S	6.8-7.2	A.S.	38 (262)	25 (172)	7.0	21.0 (14.5)	42R_b	32.0 (43.4)
	FN-0200-T	7.2-7.6	A.S.	45 (310)	30 (207)	10.5	23.0 (15.8)	51R_b	50.0 (67.8)
Fe-4.25Ni-1.0Cu-0.15C	FN-0400-R	6.4-6.8	A.S.	36 (248)	22 (152)	5.0	17.0 (11.7)	40R_b	16.0 (21.7)
	FN-0400-S	6.8-7.2	A.S.	49 (338)	30 (207)	6.0	21.0 (14.5)	60R_b	35.0 (47.5)
	FN-0400-T	7.2-7.6	A.S.	58 (400)	36 (248)	6.5	23.0 (15.8)	67R_b	50.0 (67.8)
Fe-7.0Ni-1.0Cu-0.15C	FN-0700-R	6.4-6.8	A.S.	52 (358)	30 (207)	2.5	17.0 (11.7)	60R_b	12.0 (16.3)
	FN-0700-S	6.8-7.2	A.S.	71 (489)	40 (276)	4.0	21.0 (14.5)	72R_b	21.0 (28.5)
	FN-0700-T	7.2-7.6	A.S.	85 (586)	48 (331)	6.0	23.0 (15.8)	83R_b	26.0 (35.2)
Nickel steel									
Fe-2.0Ni-1.25Cu-0.45C	FN-0205-R	6.4-6.8	A.S.	37 (255)	23 (159)	3.0	17.0 (11.7)	47R_b	10.0 (13.6)
	FN-0205-R	6.4-6.8	H.T.	82 (565)	65 (448)	0.5	17.0 (11.7)	32R_c	6.0 (8.13)
	FN-0205-S	6.8-7.2	A.S.	50 (345)	31 (214)	3.5	21.0 (14.5)	74R_b	18.0 (24.4)
	FN-0205-S	6.8-7.2	H.T.	110 (758)	88 (607)	1.0	21.0 (14.5)	42R_c	16.0 (21.7)
	FN-0205-T	7.2-7.6	A.S.	61 (421)	37 (255)	4.5	23.0 (15.8)	85R_b	32.0 (43.4)
	[illegible]	[illegible]	[illegible]	[illegible]	[illegible]	[illegible]	23.0 (15.8)	[illegible]	28.0 (38.0)

	[illegible]	[illegible]	A.S.	97 (669)	50 (345)	...	22.0 (15.1)	$90R_b$	27.0 (36.6)
Fe-2Ni-1.25Cu-0.75C	FN-0208-R	6.4-6.8	A.S.	48 (331)	30 (207)	2.0	17.0 (11.7)	$62R_b$	8.0 (10.8)
	FN-0208-R	6.4-6.8	H.T.	100 (689)	94 (648)	0.5	17.0 (11.7)	$34R_c$	6.0 (8.1)
	FN-0208-S	6.8-7.2	A.S.	65 (448)	41 (283)	3.0	21.0 (14.5)	$79R_b$	14.0 (19.0)
	FN-0208-S	6.8-7.2	H.T.	135 (931)	128 (882)	0.5	21.0 (14.5)	$45R_c$	12.0 (16.3)
	FN-0208-T	7.2-7.6	A.S.	79 (545)	50 (345)	3.5	23.0 (15.8)	$87R_b$	22.0 (29.8)
	FN-0208-T	7.2-7.6	H.T.	160 (1103)	155 (1069)	0.5	23.0 (15.8)	$47R_c$	18.0 (24.4)
Fe-4.25Ni-1.0Cu-0.45C	FN-0405-R	6.4-6.8	A.S.	45 (310)	26 (179)	3.0	17.0 (11.7)	$63R_b$	10.0 (13.6)
	FN-0405-R	6.4-6.8	H.T.	112 (772)	94 (648)	0.5	17.0 (11.7)	$27R_c$	6.0 (8.1)
	FN-0405-S	6.8-7.2	A.S.	62 (427)	35 (241)	4.5	21.0 (14.5)	$72R_b$	15.0 (20.3)
	FN-0405-S	6.8-7.2	H.T.	154 (1062)	128 (882)	1.0	21.0 (14.5)	$39R_c$	10.0 (13.6)
	FN-0405-T	7.2-7.6	A.S.	74 (570)	43 (296)	6.0	23.0 (15.8)	$80R_b$	30.0 (40.7)
	FN-0405-T	7.2-7.6	H.T.	180 (1241)	154 (1062)	1.5	23.0 (15.8)	$44R_c$	14.0 (19.0)
Fe-4.25Ni-1.0Cu-0.45C	FN-0408-R	6.4-6.8	A.S.	57 (393)	42 (290)	1.5	17.0 (11.7)	$72R_b$	6.0 (8.1)
	FN-0408-S	6.8-7.2	A.S.	77 (531)	57 (393)	3.0	21.0 (14.5)	$88R_b$	10.0 (13.6)
	FN-0408-T	7.2-7.6	A.S.	93 (641)	68 (469)	4.5	23.0 (15.8)	$95R_b$	16.0 (21.7)
Fe-7.0Ni-1.0Cu-0.45C	FN-0705-R	6.4-6.8	A.S.	54 (372)	35 (241)	2.0	17.0 (11.7)	$69R_b$	9.0 (12.2)
	FN-0705-R	6.4-6.8	H.T.	107 (703)	80 (552)	0.5	17.0 (11.7)	$24R_c$	8.0 (10.8)
	FN-0705-S	6.8-7.2	A.S.	76 (524)	48 (331)	3.5	21.0 (14.5)	$83R_b$	17.0 (23.0)
	FN-0705-S	6.8-7.2	H.T.	140 (965)	110 (758)	1.0	21.0 (14.5)	$38R_c$	15.0 (20.3)
	FN-0705-T	7.2-7.6	A.S.	90 (620)	57 (393)	5.0	23.0 (15.8)	$90R_b$	24.0 (32.5)
	FN-0705-T	7.2-7.6	H.T.	168 (1158)	130 (896)	1.5	23.0 (15.8)	$40R_c$	20.0 (27.1)
Fe-7.0Ni-1.0Cu-0.45C	FN-0708-R	6.4-6.8	A.S.	57 (393)	41 (283)	1.5	17.0 (11.7)	$75R_b$	6.0 (8.1)
	FN-0708-S	6.8-7.2	A.S.	80 (552)	55 (379)	2.5	21.0 (14.5)	$88R_b$	12.0 (16.3)
	FN-0708-T	7.2-7.6	A.S.	95 (655)	66 (455)	3.0	23.0 (15.8)	$96R_b$	16.0 (21.7)

(a) A.S.: as sintered. H.T.: heat treated. (b) ASTM E8. (c) International Nickel Co.

Table 6. Material requirements for P/M pump gear applications(14)

Application	Material code
Output pressure less than 100 psi and light service	FC-0208 type; as-sintered; 5.8-6.2 g/cm^3
Output pressure less than 1000 psi but subject to high vibration and heavy service; hardened by heat treating	FC-0208 or FN-0106; 6.4-6.8 g/cm^3
Medium high pressure (1000-1250 psi) for short intervals; hardened by heat treating	FC-0208 type; 6.2-6.5 g/cm^3
General-purpose hydraulic pumps up to 1500 psi continuous service	FN-0106 type; heat treated; 6.8-7.1 g/cm^3
General-purpose hydraulic pumps up to 2500 psi continuous service	AISI 4630 type; heat treated; 7.2-7.6 g/cm^3

carbon powders with densities of 6.7 to 6.8 g/cm^3. Another automatic transmission application is shown in Fig. 15. These are the components of variable displacement pumps typical of several models in current production. The outer rings were made from iron-carbon-nickel powder and the

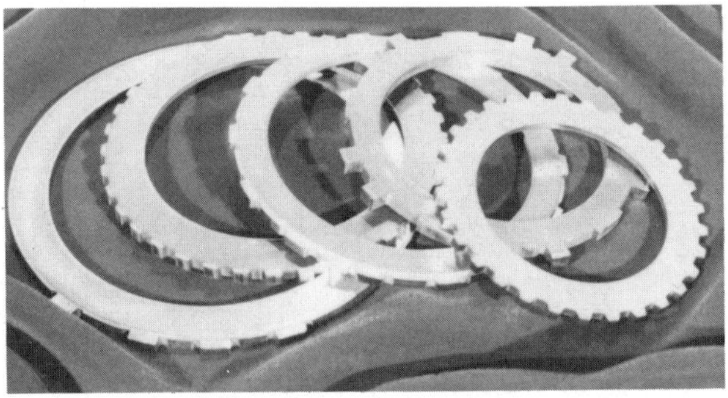

Fig. 13. P/M clutch pressure plates used in an automatic transmission(16)

inner rings were made from iron-carbon-nickel-copper powder. Both outer and inner rings have densities of 6.7 to 6.8 g/cm^3.

High-Density Ferrous Powder Metallurgy

For high-density P/M pressed and sintered parts, secondary operations such as re-pressing are commonly used. In

Fig. 14. P/M crankshaft pulley used for driving the timing belt in an automotive engine[16]

Fig. 15. P/M parts for variable displacement pumps in an automatic transmission[16]

the re-pressing process, pressed parts are presintered at a low temperature for a period long enough to burn off any lubricant. The presintered parts are then re-pressed to a higher density and sintered at a conventional temperature. Typical density levels of presintered parts are 6.8 to 7.0 g/cm^3, and this level is increased to 7.3 to 7.5 g/cm^3 after re-pressing. Consequently, mechanical properties of re-pressed P/M parts are increased, allowing them to be used for intermediate-stress applications.

Another common technique for providing P/M parts for high-stress applications is infiltration. Infiltration is a process in which interconnected pores in a P/M compact are filled with a liquid metal or alloy whose melting point is much lower than that of the compact. The driving force of infiltration is the minimization of the total surface-free energy of the system. It is necessary that the total surface-free energy of the system after infiltration, including the surface-free energy of the solid and liquid phases by themselves and the interfacial energy between solid and liquid, be lower than the total surface-free energy before infiltration. Generally, infiltration will be promoted by using an infiltrant with a low wetting angle and a low viscosity, by providing a pressure gradient and by infiltrating P/M parts

Table 7. Compositions and typi

MPIF designation	Composition(a), % C	Cu	Condition (b)	Tensile strength MPa	ksi
FX-1005-T. . .	0.3-0.6	8.0-14.9	AS	570	83
			HT	830	120
FX-1008-T. . .	0.6-1.0	8.0-14.9	AS	620	90
			HT	895	130
FX-2000-T. . .	0.3 max	15.0-25.0	AS	450	65
FX-2005-T. . .	0.3-0.6	15.0-25.0	AS	515	75
			HT	790	115
FX-2008-T. . .	0.6-1.0	15.0-25.0	AS	585	85
			HT	860	125

(a) Remainder Fe. (b) AS: as sintered. HT: heat treated (typic notched Charpy test.

with large interconnecting pores. The following advantages of infiltration of P/M parts were discussed by Durdaller[17]:

- Increased mechanical properties such as tensile strength, hardness, impact resistance and fatigue strength
- Uniform density in parts
- Production of high-density parts (up to 7.2 g/cm^3) by a single pressing and sintering operation
- Elimination or sealing of surface pores for secondary operations such as pickling and plating
- Production of P/M parts with selective density variations
- Assembly of multiple parts by sintering the individual pieces together and bonding the pieces into one part through common infiltration.

Coppers and copper-base alloys are commonly used for infiltration of ferrous P/M parts because of their relatively low melting points. Infiltration of ferrous parts can be accomplished by one of the following processes:

- P/M compacts and infiltrants are placed together in a sintering furnace and infiltration takes place during the sintering process.
- P/M compacts are sintered without infiltrant. Sintered P/M compacts are then placed in a furnace with an

operties of P/M infiltrated steels[12]

Yield strength MPa	ksi	Elongation in 25 mm or 1 in., %	Impact energy(c) J	ft·lb	Apparent hardness	Elastic modulus GPa	10^6 psi
440	64	4.0	19	14	75 HRB	135	20
740	107	1.0	9.5	7.0	35 HRC	135	20
515	75	2.5	16	12	80 HRB	135	20
725	105	60.5	9.5	7.0	40 HRC	135	20
...	...	1.0	20	15	60 HRB	...	...
345	50	1.5	12.9	9.5	75 HRB	125	18
655	95	<0.5	8.1	6.0	30 HRC	125	18
515	75	1.0	14	10	80 HRB	125	18
740	107	<0.5	6.8	5.0	42 HRC	125	18

tenitized at 870°C, oil quenched, and tempered 1 h at 200°C). (c) Un-

infiltrant and heated. Carbon-free iron powder parts are not normally infiltrated with copper because of excessive growth of the parts during infiltration. It has been estimated[17] that this growth is on the order of 2 to 3% for carbon-free iron powder but that addition of 1.25% graphite decreases the dimensional change to about 0.1%. Compositions and typical properties of P/M infiltrated steels are given in Table 7 on the preceding two pages.

Part II: Fully Dense Parts and Their Applications

B. Lynn Ferguson
Consultant
Deformation Control Technology
(Formerly with TRW Materials Technology)

P/M Forging Technology

The steps in the P/M forging process are:

1 Consolidation of preform
2 Sintering of preform
3 Preheating of preform and forging tooling (for hot or warm forging)
4 Forging
5 Part finishing.

The first two steps comprise preform production and are discussed in the next section. Subsequent sections discuss preform design, the actual forging step, the effects of forging variables on mechanical properties and performances, and specific applications.

Preform Production

A preform is a porous compact that has been pressed and sintered to a desired shape and density to facilitate forging. The methods of compacting and sintering iron and steel powder preforms are exactly those utilized to make press-sinter parts. Water-atomized alloy powder is the preferred starting material, but blended powder mixes are also acceptable. For nonferrous powder forging, premixes may be preferable to prealloyed powder. The key to preform production is the ability to cold compact the powder into the desired shape.

The preform shape may be either simple or complex, depending on the nature of the part to be forged, as will be discussed in the following section. Simple shapes will most likely be deformed into more complex part shapes, which means that the surface finish and dimensional control of the preform are less critical than the preform weight. For such preforms, both hard die compaction and cold isostatic pressing are possible preform production methods. For preforms of complex shape, which typically are subjected to limited deformation because the final shape is similar to the preform shape, hard die tooling should be used since preform surface finish and dimensions are now important. These are not strict rules. Each part must be examined individually to determine which type of compaction should be used, with economics usually dictating the final decision.

The important features of the preform are its shape, weight and density. More will be discussed about these features in the section on preform design. It should be emphasized that the preform production method selected must be capable of producing the desired values of these variables.

The sintering step of this process serves the same purpose as in press-sinter part production. Metal oxides must be reduced, carbon should be diffused into the alloy, and chemical homogenization should occur.

Preform Design

Selection of a preform for forging to final part shape is the critical step in P/M forging. Preform shape and density directly affect the mechanical properties, particularly toughness and fatigue resistance, that are achieved in the part. Porosity in the preform drastically reduces the resistance to fracture of the material, which means that surface cracking of the deforming preform can be a serious problem. Historically, successful preform designs have been developed by an iterative trial and error procedure, using prior experience to determine the initial shape. Recent developments have been reported which employed CAD (computer-aided design) for preform design[18,19].

To understand the concept of preform design, first consider forging of a simple solid cylinder. The preform may range from a re-pressing-type preform, where deformation is limited to axial collapse as porosity is eliminated, to a high aspect ratio preform, where upsetting first occurs, followed by re-pressing as die side wall contact occurs. These two processes are shown schematically in Fig. 16. Because upsetting introduces tensile strains to the bulging surface of the preform, the workability of the material becomes important.

A method of characterizing workability where free surface fracture is predominant, as is the case during upsetting, has been developed for wrought materials[20], and has been successfully applied to porous preforms[22,23]. A schematic in Fig. 17 shows a typical workability line and how

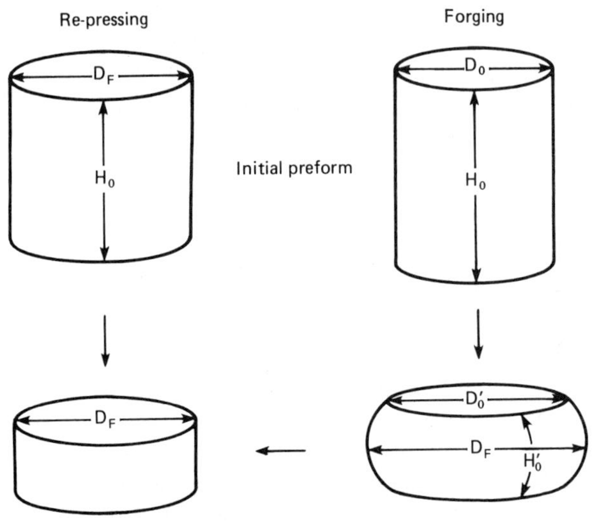

Fig. 16. Geometries of preforms used for re-pressing and forging of a cylindrical slug. In re-pressing, only the height, H_0, changes; in forging, both diameter and height change.

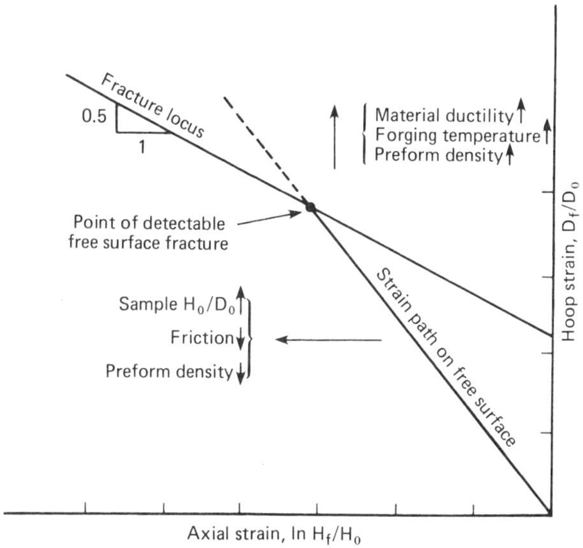

Fig. 17. Qualitative effects of forging variables on formability of porous preforms during hot forging[24]

process variables affect the line location[24]. The line has a slope of −0.5, and workability improves as the plane strain intercept (y-axis intercept) value increases. For a given material, workability can be improved through temperature adjustment or density changes. Alternatively, surface fracture problems may be avoided by changing the preform aspect ratio or the forging lubricant.

For a multilevel part, preform design becomes more critical. Re-pressing is no longer a simple task of matching up the final shape and preform shape through an adjustment for porosity as it is for symmetrical single-level parts. An example of this is shown in Fig. 18, where the forging operation depicted is re-pressing a two-level part from a shaped preform of uniform density. In Fig. 18(a), a solid top punch makes contact on the two levels simultaneously. As the punch moves down, axial collapse due to elimination of porosity occurs. In Fig. 18(b), full density has been reached in the thin section, but porosity still remains in the thick section. Continued punch travel forces material to

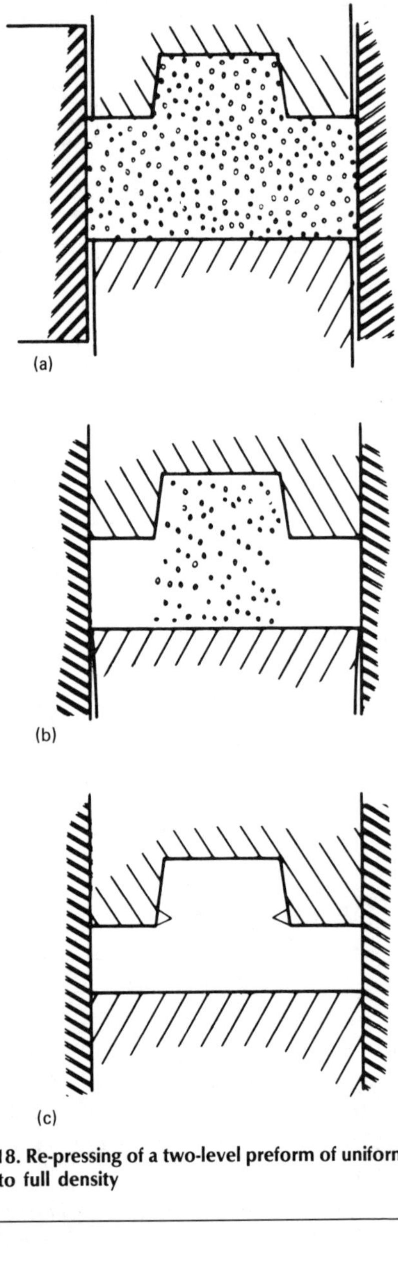

(a)

(b)

(c)

Fig. 18. Re-pressing of a two-level preform of uniform density to full density

flow from the thin section into the thick section. The preform corner rotates and moves up onto the thick section wall, as shown in Fig. 18(c). At the point of full density, the corner is collapsed to form a forging defect. An example of this is shown in Fig. 19. Several techniques are available to overcome this problem. The two best procedures are: (*a*) production of a preform with nonuniform density such that full density is reached simultaneously between preform sections; or (*b*) the use of split punches in the forging tooling to accommodate local flow at junctions.

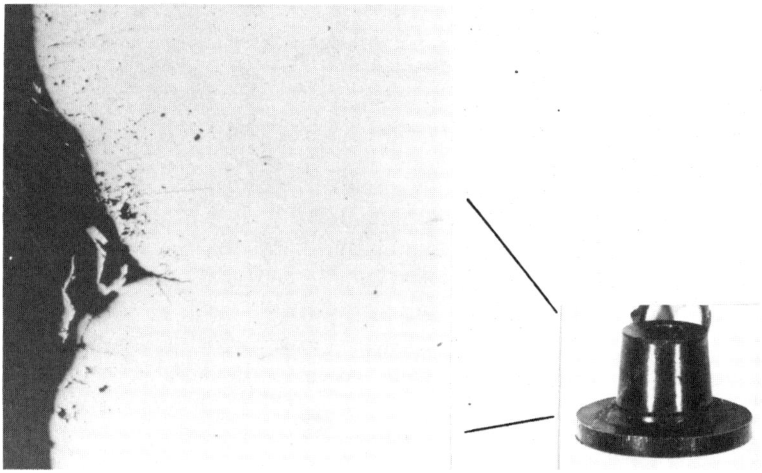

Fig. 19. Defect caused by internal metal flow in a two-level preform during re-pressing.[25] Preform corner rotated and moved onto hub as metal flow occurred from flange into hub.

For simple-shape preforms, the workability analysis briefly presented applies as well as workability for two added problem areas. These are internal cavitation due to directed metal flow and surface fracture at die contact areas due to tensile stresses caused by friction gradients[23,26,27].

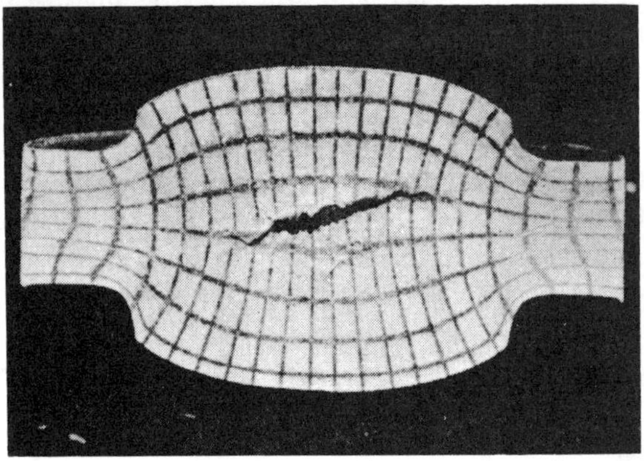

Fig. 20. Internal bursting during simultaneous forward and backward extrusion[27]

Figure 20 shows a defect caused by metal flow in opposing directions, generating a cavity. Proper preform design can minimize this problem of internal bursting. Figure 21 shows an example of a die-contacted surface fracture that was caused by an artificially large friction gradient. Figure 22 is an actual example of die contact fracture which, in this case, was caused by a change in the contact surface geometry.

An example of the effect of preform geometry is taken from Downey and Kuhn[21]. Figure 23 shows four possible preforms which could be forged to produce an axisymmetric part having cup and hub sections. Preform geometries in parts (b), (c) and (d) result in defective forgings due to cracking along the center bore of the hub, the internal cup surface or the top of the cup. Preform (a) resulted in a defect-free forging. Because preform design is a key ingredient for successful P/M forging, and because of the complexity of this issue, it is not possible to develop the subject to the necessary depth in this review. Suggested references on this subject are Ref. 18, 22, 23, 26, 27 and 28.

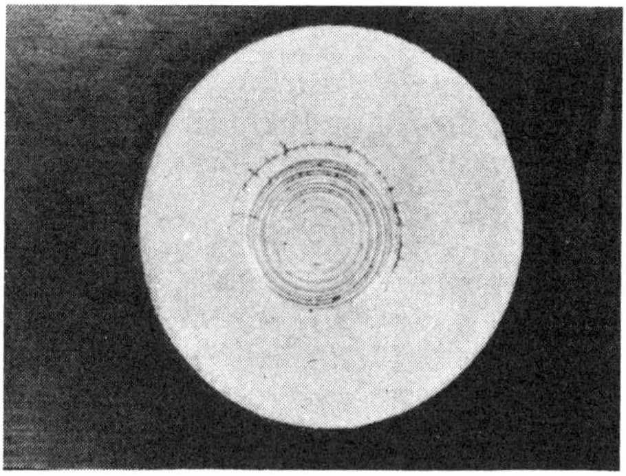

(a)

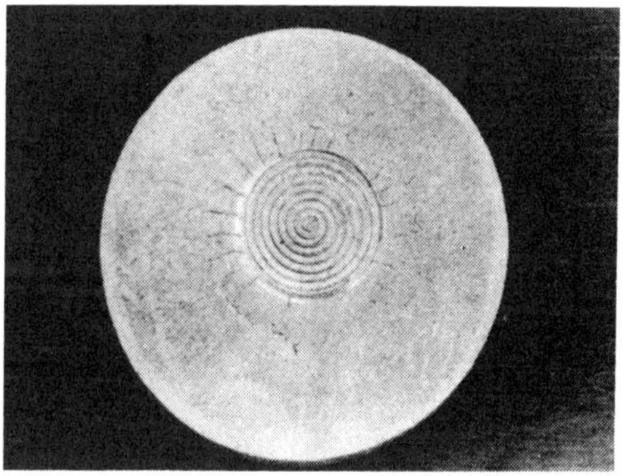

(b)

Both radial and circumferential cracks are visible in the vicinity of the transition (a and b). In addition to the cracks, interconnected pores were opened in the smooth outer region.

Fig. 21. Top views of disks compressed between dies having a grooved central region and a smooth outer region with grease lubrication[27]

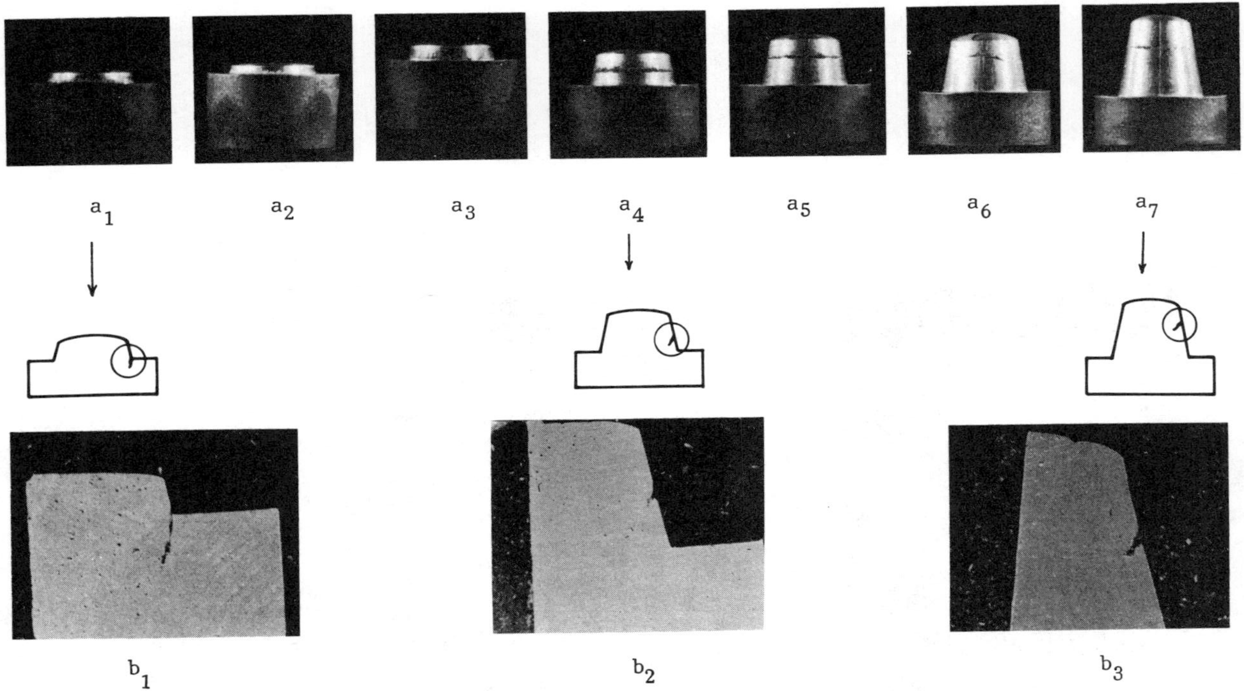

Fracture occurs only once in the early stage of deformation and moves up the hub surface (a_1-a_7). The angle with respect to the forging direction increases (b_1-b_3) as deformation continues.

Fig. 22. Change of contact surface fracture (a) and cross-sectional view (b) of its angular change[27]

The Forging Step

Once a preform has been designed and produced, forging can begin. Critical variables are:

- Forging temperature – preform and die
- Transfer of preform from preheat furnace to press
- Ejection temperature
- Lubrication.

Die design and the forging press to be used are equally important considerations. For steel powders, the preform

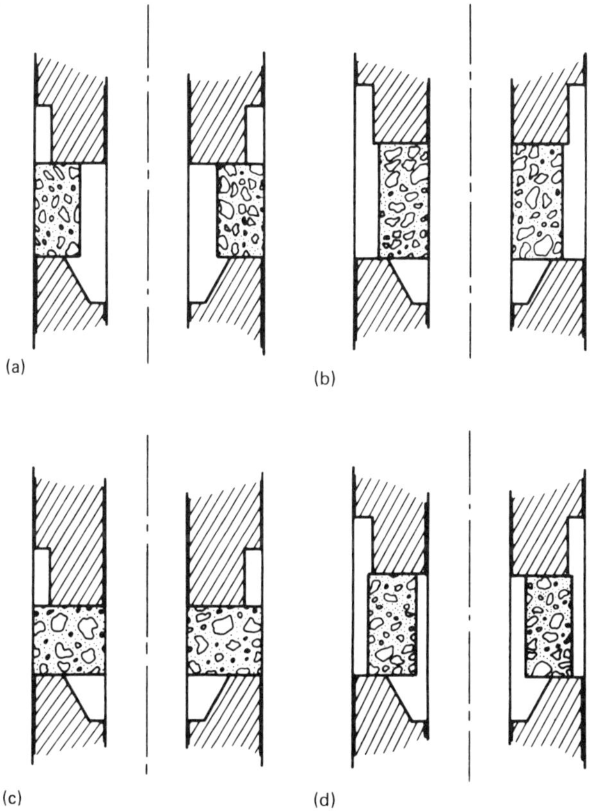

Fig. 23. Schematic illustration of the four ring-preform options for production of a complex part having cup and hub sections[21]

is typically coated with a graphite lubricant, heated under a protective atmosphere to the forging temperature, transferred in air to the forging die and forged in one blow to full density. The die is heated to an arbitrary value to minimize die chill and also is lubricated with graphite. The application of lubricant usually aids die cooling in production operations.

Die design involves the design of a confined die set, including die materials selection, to produce a flashless part. Die dimensions must take into account the preform temperature, the die preheat temperature, the forging load, the elastic strain of the die, the elastic/plastic strain of the forging, the temperature upon ejection, the elastic strain of the forging upon ejection and the contraction of the forging during cooling. Part dimensional tolerances can be met only when these features are accounted for. Once the die dimensions have been selected, there is some flexibility in dimensional control of the forged part in that the preheat temperatures can be adjusted. For example, if the forging is undersize for a given set of forging conditions, a lower preform preheat temperature and/or a higher die preheat temperature can be used to produce a larger part. Similarly, if the forging is oversize, the preform preheat temperature should be raised and/or the die temperature lowered to bring the part into proper dimensions. A higher preform temperature results in a higher ejection temperature for the forging. It will thus shrink a greater amount during cooling than would a forging with a lower ejection temperature. An increase in die temperature expands the die cavity to increase the size of the forging.

Transfer of the heated preform to the forging die must be rapid, because internal oxidation of the preform due to the presence of pores results in an unacceptable forging. During preheating, the protective atmosphere of the furnace circulates through the preform. Some atmosphere remains with the preform during transfer and exudes from the pores to protect the preform. Depending on preheat temperature and preform density, there is a finite time span, generally ranging from 4 to 20 seconds, during which this protection is effective(29,30). Forging must occur prior to the end of this period.

The preceding paragraphs centered on hot forging, but warm and cold forging of ferrous P/M preforms are also suitable deformation processes. The same principles apply, except that preform workability is reduced as the working temperature is reduced, and part dimensional sizing control becomes even tighter. Shapes such as bearing cups which can be formed such that only compressive strains occur throughout the preform have been successfully cold forged[31].

Effects of Forging Variables on Mechanical Properties

For P/M forged parts, final mechanical properties depend on two main features: sintering conditions and deformation during part shaping. By control of these two features, the mechanical properties of a part can be optimized.

First consider the effect of porosity on mechanical properties. Strength, ductility, toughness and fatigue resistance increase as porosity is eliminated. The latter three properties are especially sensitive to porosity and increase dramatically as the last traces of porosity are removed. This effect is shown for toughness in Fig. 24[30]. The elimination of porosity in highly stressed regions of a part is essential for achievement of property levels comparable to wrought property levels.

Furthermore, these sensitive properties improve as the level of upsetting is increased during the forging process. Figure 25 shows improvement in Charpy V-notch impact energy absorbed values as the deformation level increases[32]. Figure 26 shows a similar effect for fatigue resistance, although the cyclic stress state also influences fatigue behavior[32]. An interesting feature of P/M forged parts is the fact that deformation does not significantly affect through-thickness properties as it does detrimentally for wrought material[33]. For re-pressed parts, through-thickness toughness is slightly lower than longitudinal toughness. Upsetting increases longitudinal toughness while toughness in the through-thickness direction remains at a relatively constant level.

Figure 25 shows that sintering variables affect toughness[32]. In this case, the higher sintering temperature results in a lower oxide level. Particle/particle bonding

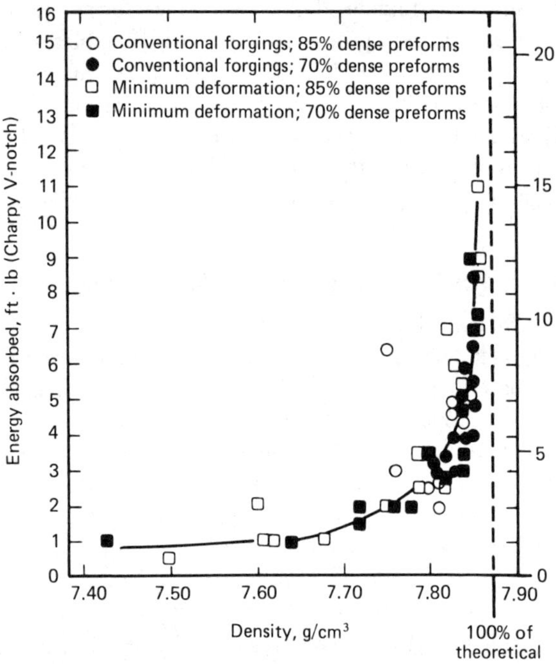

Fig. 24. Charpy V-notch strength of P/M forgings as a function of density[30]

during forging is enhanced because the internal surfaces are cleaner. The result is higher toughness at lower levels of flow for material sintered at 2300°F rather than 2050°F. Toughness in the short transverse direction is also improved by high-temperature sintering[25].

The freedom to achieve property levels by adjusting sintering conditions or the preform shape, which controls the deformation level, gives the designer and partsmaker a wide processing window. In cases where deformation must be minimized because of part shape constraints, re-pressing can be coupled with high-temperature sintering to maximize properties for that part. If flow can be accommodated, high dynamic properties can be achieved at normal sintering temperatures by maximizing the amount of metal flow needed to form the part. Suggested readings

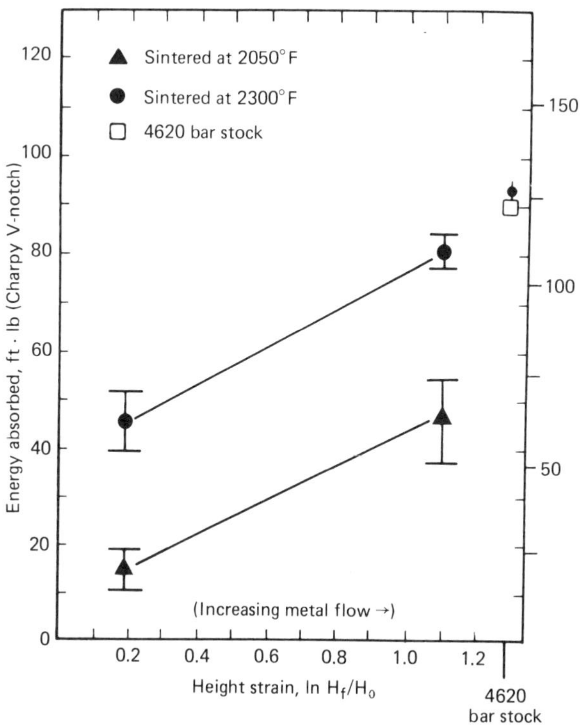

Fig. 25. Toughness of 4620 P/M forgings as a function of flow during forging and of sintering temperature[32]

on processing effects on mechanical properties of P/M forgings include Ref. 25, 29, 30, 32, 33, 34 and 35.

Applications

This section describes a few representative P/M forged parts that illustrate some of the ideas presented above.

Part 1: Machine Gun Accelerator[30]. A program was sponsored at TRW Materials Technology by Rock Island Arsenal to develop a manufacturing process that would lower current production costs for a machine gun accelerator. The current process included hot forging of 4340 steel bar stock to a rough shape by six forging opera-

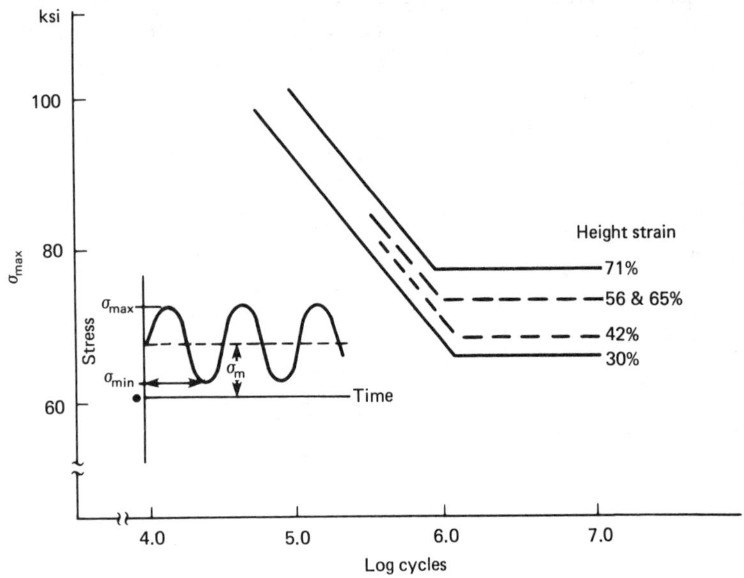

Fig. 26. Axial fatigue of P/M forged 4620 steel as a function of height strain during forging[32]

tions, followed by twenty-seven machining operations to arrive at a finished part. P/M forging was selected as a cost-cutting process because it could forge a near-net part in a single blow.

The preform, forging and finished accelerator are shown in Fig. 27. Because of the large surface area-to-volume ratio and the height-to-thickness ratio of the accelerator arms, a preform involving minimum deformation to achieve the final shape was designed. For this type of preform, metal flow between part sections must be minimal. Therefore, density levels within the preform had to be varied to satisfy this requirement. Split punch tooling, as shown in Fig. 28, was utilized to compact this preform. During forging, the stroke length of the upper punch resulted in the achievement of theoretical density in all sections simultaneously.

Water-atomized 4600 steel powder was blended with graphite and compacted in the split punch tooling at a

Fig. 27. P/M forged machine gun accelerator: preform, forging and finished part(30)

Fig. 28. Split-punch tooling used for compaction of minimum-deformation-type preform for machine gun accelerator(30)

pressure of 30 tsi. Following compaction, the preform was sintered at 2200°F for 1 hour in an atmosphere of dry hydrogen plus 0.5 vol % methane. The final carbon level was 0.40 wt % and the oxygen level was reduced to the range 150 to 200 ppm. The preform was preheated to 2200°F and forged in one blow in a confined die heated to 400°F. Only seven machining steps were required to produce a finished part.

Tensile properties of 4640 P/M forged material and wrought 4640 are compared at various hardness levels in Fig. 29. The data for P/M forgings (re-pressed) are similar

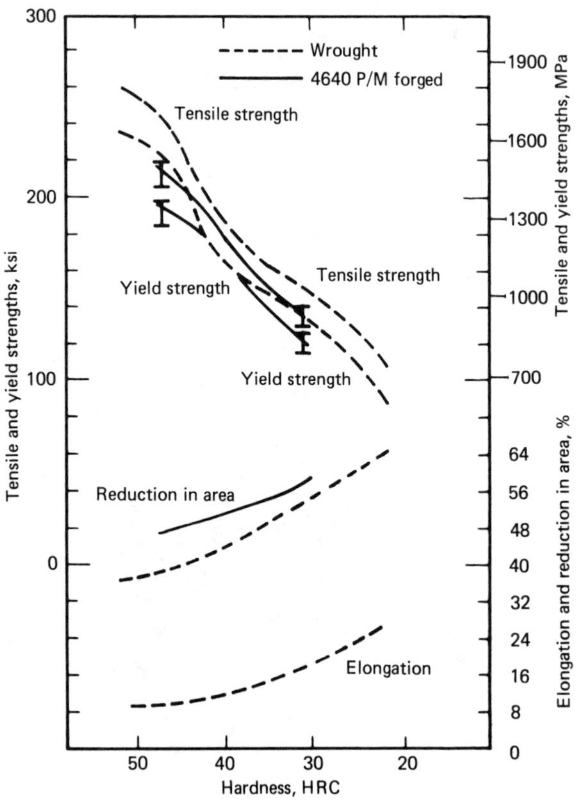

Fig. 29. Tensile properties of P/M forged 4640 steel and typical properties of wrought 4640 steel[30]

to wrought data. Fatigue data for P/M forged 4640 are shown in Fig. 30, and these data fall within the scatter band for 4340 steel. The most impressive statistic is that the P/M forged parts passed the Army ambient and low temperature firing endurance tests. Based on the satisfactory performance of the accelerator, the Army established specification MIL-F-45961 which authorizes the procurement of twenty-six machine gun components as P/M forged parts.

The cost analysis of the P/M forging process is given in Table 8 for a projected die life of 50,000 pieces. Due to reduced machining costs and more efficient utilization of raw material, P/M forging reduces part cost by 50% compared with that of a conventionally forged/machined part.

Part 2: Differential Gear Set for Army Truck[35,36]. The Tank Automotive Command (TACOM) of the U.S. Army sponsored a program at TRW Materials Technology to develop a manufacturing process for P/M forging of differential gears for light army trucks[36]. The differential gear set consisted of two side gears and two mating pinions, as

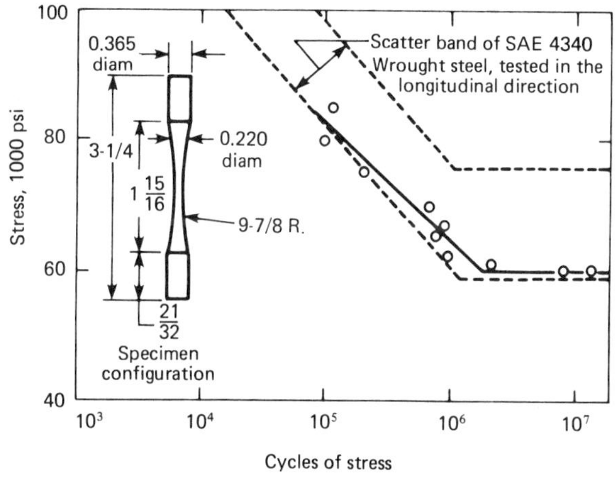

Fig. 30. R. R. Moore fatigue curve for P/M forged 4640 steel hardened and drawn to 33 HRC and a yield strength of 138,000 psi[30]

Table 8. Cost breakdown for conventional forging vs P/M forging[29]

Operation	Standard time per piece, h	Cost per piece	Setup time, h
Conventional process			
Heat and forge	0.0670	$1.68	3.0
Trim	0.0083	$0.21	1.0
Coin	0.0083	$0.21	1.0
Heat treat(a)	0.1600	$4.00	. . .
Sandblast	0.0300	$0.75	. . .
Machine (27 steps)	0.8012	$20.00	60.3
Finish	0.1882	$4.70	1.1
Material	. . .	$1.50	. . .
Total	1.2630	$33.05	66.4
Minimum deformation (P/M) process			
Compact	0.0200	$0.50	4.0
Sinter	0.0025	$0.06	. . .
Forge	0.0200	$0.50	3.0
Heat treat(a)	0.0800	$2.00	. . .
Sandblast	0.0300	$0.75	. . .
Machine (7 steps)	0.2699	$6.75	16.6
Finish	0.1882	$4.70	1.1
Material	. . .	$0.26	. . .
Total	0.6106	$15.52	24.7

(a) Cost of heat treating is proportional to weight of unmachined part.

shown in Fig. 31. The gear set shown in this figure was forged to net teeth from porous preforms.

Because of the axisymmetric shapes of these gears and the need to achieve a high level of toughness, preforms of simple shape were compacted. The preform geometries are depicted in Fig. 32. Water-atomized 4600 steel powder was blended with graphite and compacted in hard tooling at 30 tsi. The preforms were sintered at 2200°F for one hour in dry hydrogen plus 0.5 vol % methane. The carbon level of the sintered preforms was 0.20 wt %, and the oxygen level ranged from 150 to 200 ppm. The preforms were forged at 2200°F in one blow. The pinions were forged to net size, with no finish machining being required except for a chamfer on one edge. The side gears were formed by

Fig. 31. Set of differential side gears and mating pinions as P/M forged and finished(36)

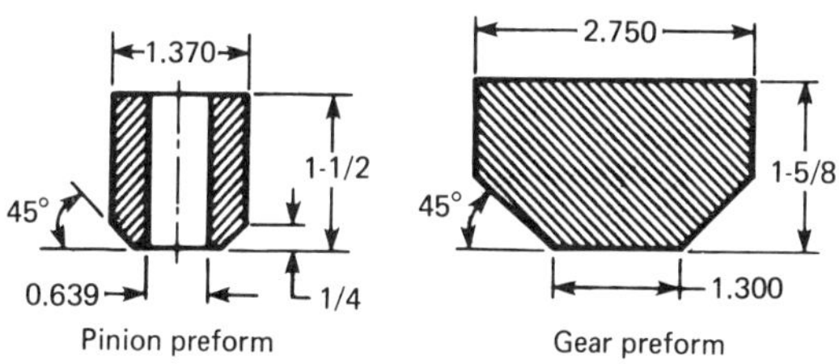

Fig. 32. Gear and pinion preform configurations(36)

a combination extrusion/forging operation in one blow. Side gears required machining on the shafts and back faces, but the teeth were forged to net size. After forging, the side gears and pinions were carburized and hardened.

Preform design was critical for these gears. At the forging temperature of 2200°F, preform shape was not critical, at least within the range of shapes examined. Preform density was a major factor as initial densities below 80% of theoretical resulted in tooth cracking. Preforms with densities above 80% possessed sufficient workability to flow into the tooth shape without cracking.

A total of 300 gear sets was supplied to TACOM for field testing[37]. The gears successfully passed this evaluation, and the Army plans to include P/M forging as an optional method for producing small gears.

The economic benefits of these gears include efficient material utilization and elimination of costly gear tooth cutting. In comparison with gears conventionally machined from bar stock, P/M forged gears can result in cost savings of up to 75%, as shown in Fig. 33.

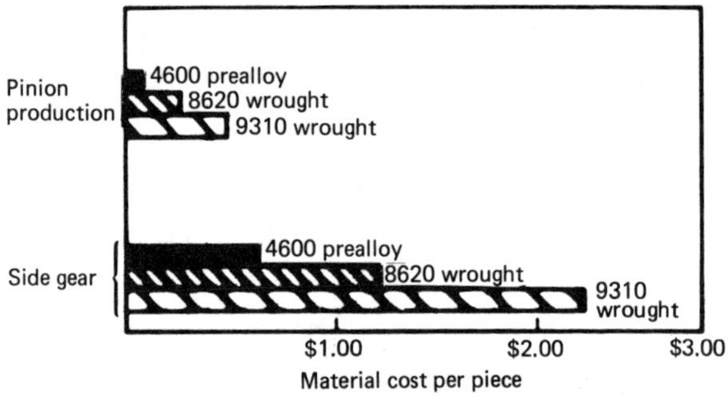

Fig. 33. Comparison of gear production material costs for P/M forging vs conventional processes[36]

A spinoff of this study illustrates another advantage of P/M forging[36]. A composite side gear was fabricated from a gear made of 4620 powder and a shaft of AISI 4620 bar stock. The sections can be assembled during either the sintering step or the forging step, resulting in a metallurgical bond. The steps used in the composite P/M forging process are shown in Fig. 34. A metallographic section of the gear/shaft bond is shown in Fig. 35. Mechanical tests revealed that the composite part was not subject to torsional failure at the bond. This process allows the designer flexibility in tailoring the properties of the gear to best meet requirements and/or to minimize costs.

Part 3: Guide Fins[38]. Another TRW P/M forging program examined the production of a tail fin for a Navy

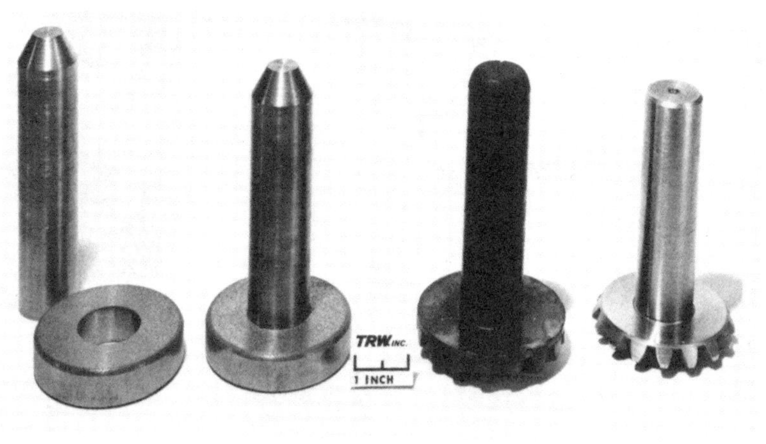

Fig. 34. Process sequence used in forging a composite side gear(36)

guided projectile. The function of the tail fin is to control the roll rate of the projectile after launch. To meet this aerodynamic requirement, the fin has rigid geometrical requirements for flatness, twist, symmetry and parallelism of opposing surfaces.

A preform requiring minimum deformation was designed for production of the P/M forged tail fin. The processing sequence is illustrated in Fig. 36. The fins are forged to near-net shape, with the airfoil surfaces being forged to net dimensions. The only required machining operations are jig boring of the mounting hole in the boss, and milling of the boss seating surface and of the latch key configuration at the free end of the fin. Grade 4640 steel powder was selected for this fin in order to meet the minimum yield strength level of 170,000 psi.

Design requirements are shown in Fig. 37, with parallelism of opposing faces, flatness of airfoil surfaces and perpendicularity of mounting hole to midplane of fin being particularly critical. To meet these stringent specifications, a process involving die quenching and hot die coining/tempering was developed. After being forged, the fins were austenitized at 1700°F, quenched to 1000°F in the forging

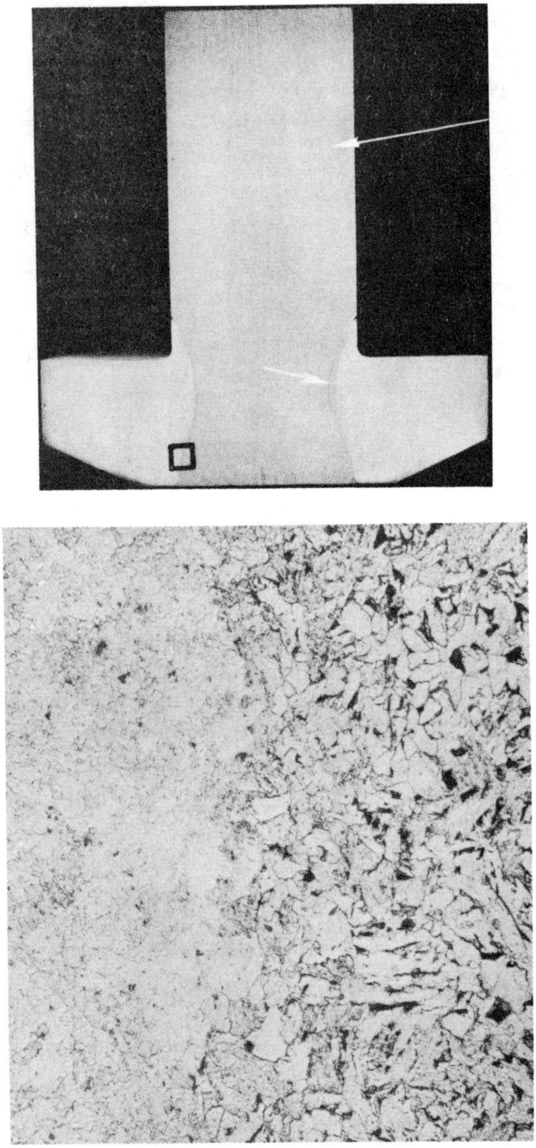

Fig. 35. Photomicrograph, at 250×, of bond between wrought shaft (top arrow) and P/M forged gear section (bottom arrow) in a composite side gear[36]

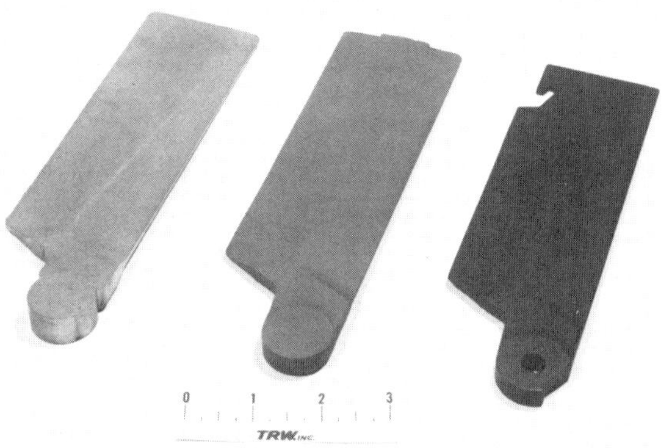

Fig. 36. Tail fin for a Navy 5-in. guided projectile, showing the 4640 P/M preform (left), the as-forged part (center) and the finish machined and black oxide coated fin (right)[38]

dies and held for 10 seconds, and then oil quenched. Next, the fins were heated at 925°F for 15 minutes and then coined in dies heated to 925°F under a load of 20 tsi for a dwell time of 1 minute. The tempered fins were slow cooled to ambient temperature. Table 9 presents tolerance ranges of critical values of oil quenched and coined/tempered fins. The Navy has been supplied with a number of fin sets for testing.

The reason for discussing this part, which presently is produced only in the laboratory, is to illustrate the versatility of P/M forging. The process can be modified to meet specification requirements such as exacting dimensions over a large plan area. The ability to forge net surfaces so that costly machining operations and fixturing can be eliminated offers economic justification for P/M forging of many parts such as these fins.

Part 4: P/M Forged Bearings[31,39]. Bearings produced by forging of porous preforms are perhaps the most successful example of ferrous P/M forging. Federal Mogul, using its patented Sinta-Forge and Sinta-Carb processes, produces a line of P/M forged components which includes

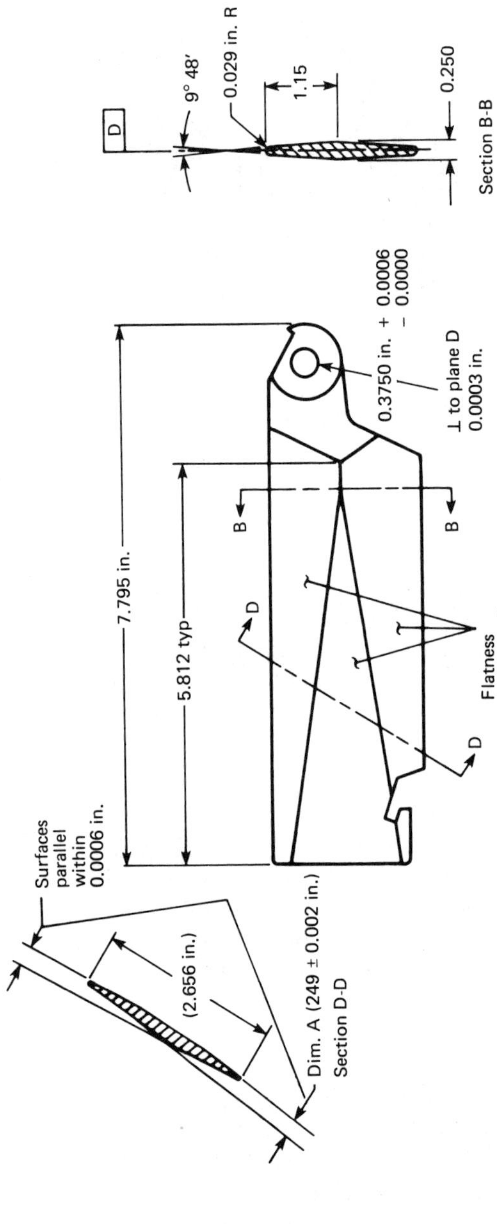

Fig. 37. Design requirements for 5-in. guided projectile fin[38]

Table 9. Characteristics of oil-quenched and coin-tempered fins[37]

Condition	Parallelism, mils Range	Mean	1 Sigma	Flatness, mils Range	Mean	1 Sigma	Hardness, HRC
As-oil-quenched	1.5-18.0	7.3	6.1	3.0-21.0	10.7	5.3	58-60
Coin-tempered	0.8-7.3	3.2	1.8	1.0-5.0	2.4	0.9	40-42

bearing races, clutch components, parts with cam surfaces, and splines and gears[39]. These parts are shown in Fig. 38. Of interest is the fact that these parts have complex geometries from the standpoint of machining, have large bores with relatively thin ring volumes of material, and are subjected to high Hertzian stresses. For hot forged P/M bearing components, both cups and cones were found to have superior rolling contact fatigue lives in comparison to components machined from wrought bar, as shown in Fig. 39[39]. P/M forged cups exhibited lives 3½ to 4 times

Fig. 38. Sinta-Forge production parts[39]

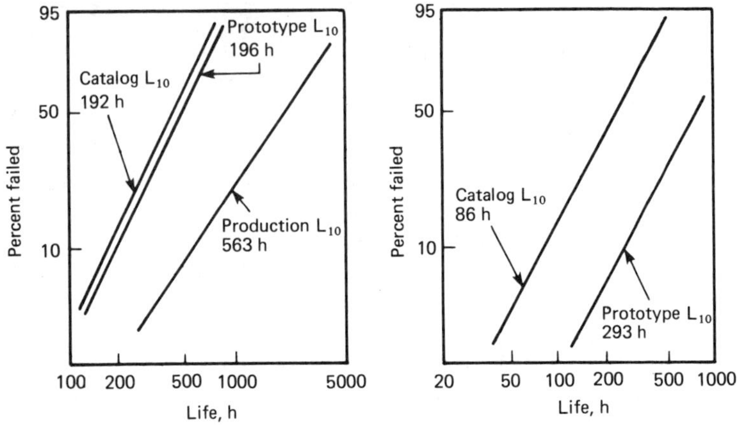

Fig. 39. Weibull plots of rolling contact fatigue tests on bearing races.[39] (Left) Tapered cone. (Right) Tapered cup.

longer than those of wrought steel cups. A cost savings of 50% was realized by substituting the P/M forged components for machined components.

Cold forging has also been successfully applied to bearing production[31]. A preform, a section of which is depicted in Fig. 40, was pressed, sintered and cold forged in one blow to greater than 98% of theoretical density. The deformation during cold forging involved limited back extrusion, significant lateral flow and final re-pressing to complete die fill. The preform was designed to keep the stress state of the deforming surfaces in safe regions on the workability map for 4600 powder preforms with an initial density 80% of theoretical. P/M cold forging resulted in closer tolerances, better surface finish (especially in ball path), no heat treat distortion, greater flexibility in material selection and lower forming loads than are obtained with conventional cold forging of tube stock. P/M ball races exceeded the performance of screw machined cups by virtue of their superior ballpath surface finish.

P/M Specialty Steels

An area of increasing importance in ferrous metallurgy is the application of P/M technology to production of

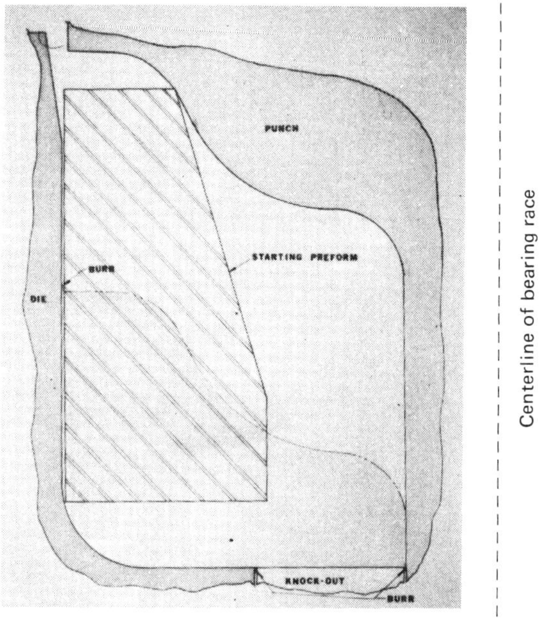

Fig. 40. Illustration showing relationship of preform to die set[31]

high-alloy steels. The advantages of P/M in achieving a uniform microstructure composed of fine constituents and in the ability to utilize higher alloy levels are being employed to produce improved tool steels, specialty steels and ferrous superalloys. The purpose of this section is to briefly introduce the use of P/M in this field.

P/M Tool Steels

Tool steels produced from powder are rapidly becoming a significant fraction of the tool steel market[40-45]. Inherent advantages of P/M become pronounced for these high alloy content steels. First, a uniform dispersion of fine second-phase constituents, including MC carbides, is obtained in P/M material. Cast/wrought tool steels are subject to carbide banding and large carbide formations. Figure 41 shows a comparison of carbide size distributions in conventional and P/M tool steels. P/M tool steel has both a finer mean carbide size and a narrower range of sizes than

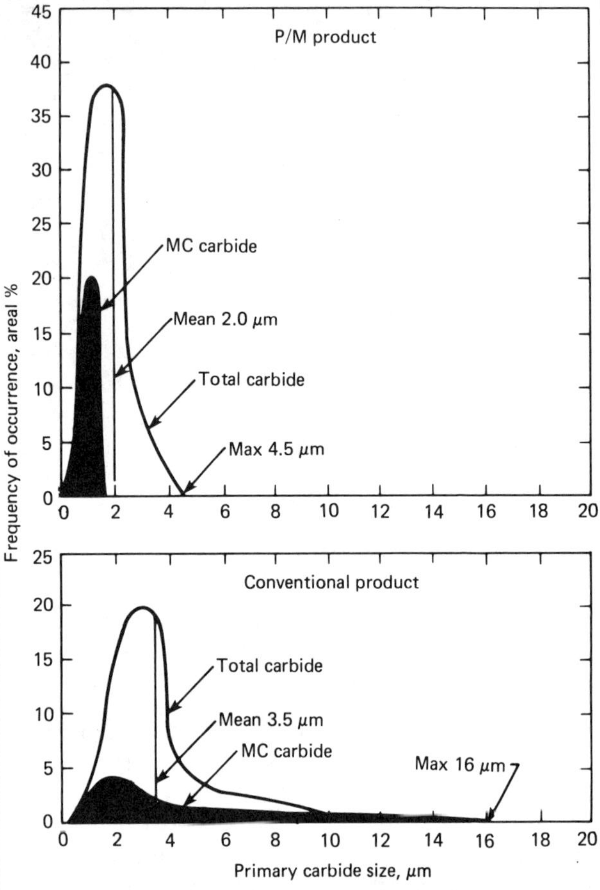

Fig. 41. Primary carbide size distributions in P/M and conventional M2S (1.0% C) bars[40]

conventional tool steel. The metallographic comparisons in Fig. 42 illustrate the difference in size and also the difference in uniformity[40].

The achievement of a refined microstructure in P/M tool steels results in better grindability and machinability than those of conventional tool steel of the same composition[40,45]. Grindability is especially important because all

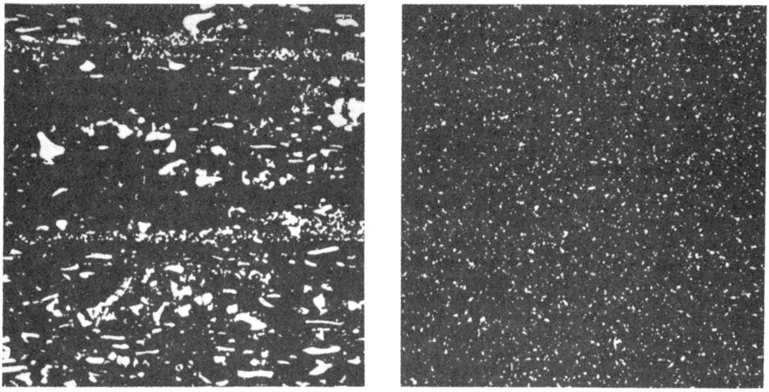

Fig. 42. Comparison of typical carbides in wrought (left) and P/M (right) T-15 steel. (Courtesy of Crucible Steel)

tools get ground during manufacturing. Figure 43 shows that P/M alloys have significantly higher grinding ratios than conventional alloys(40). This means that more tools can be ground between wheel redressings and that grinding wheels last longer.

P/M tool steel hardens more uniformly than conventional tool steel, again due to its more refined microstructure(40,45). Fine carbides go into solution more rapidly than large carbides during austenitization so that shorter times can be used and all carbides can be dissolved. Quenching and tempering result in less distortion in P/M material than in conventional material. Figure 44 shows that out-of-roundness distortion is lower in P/M material, probably due to the absence of the segregation which is present in the ingot and the absence of the box/diamond working to which conventional ingots are subjected.

Cutting performance of P/M tools and conventional tools is very similar for continuous cutting, although the refined structure of P/M tool steel results in more uniform tool wear(40,45). For interrupted cutting, however, P/M tools are clearly superior to conventional tools. Because of the refined microstructure and uniformity of carbide size, P/M

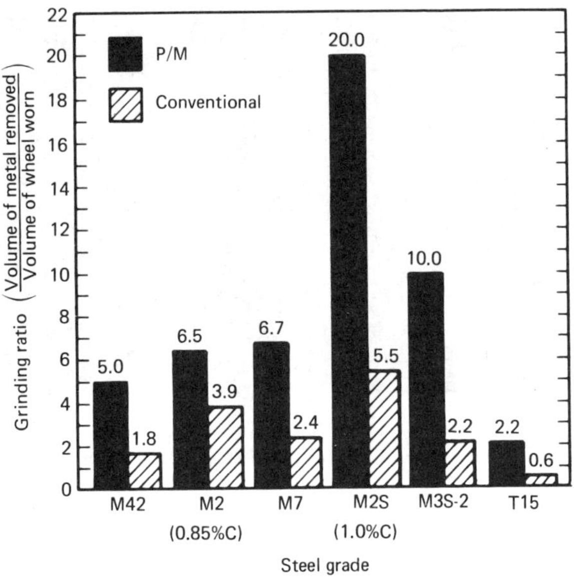

Test conditions: workpiece, 2-in.-diam by 10-in.-long bar; grinding wheel, 12-in.-OD A-60-J6-V; workpiece speed, 120 sfpm; grinding wheel speed, 6500 sfpm; infeed, 0.001 in./pass; traverse, 20 in./min; grinding fluid, water-soluble oil (heavy duty) diluted 40:1

Fig. 43. Results of grindability tests in cylindrical traverse grinding of P/M and conventional high speed steels in the fully hardened condition[40]

alloys are tougher than conventional alloys and can withstand the impact loading of intermittent cutting better than conventional alloys.

P/M techniques provide an avenue for improvement of wear resistance. Aside from microstructural refinement and toughness, P/M allows unique chemical compositions to be utilized. Tool steels of high carbide content can be produced and can be used for manufacture of various products in net or near-net form. For example, Crucible's CPM 10V can be produced only by P/M techniques because of its high vanadium carbide content[46]. P/M processing circumvents the casting and/or workability problems inherent in such high-alloy materials.

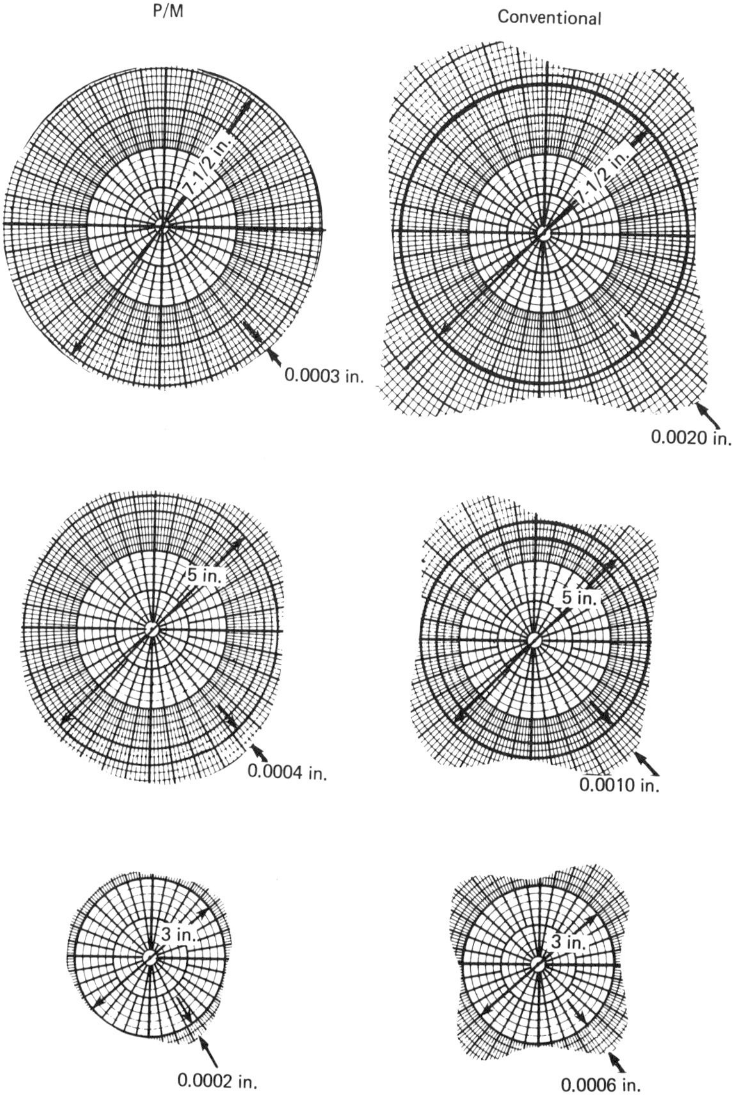

Heat treatment: preheated at 1550°F, austenitized at 2200°F, quenched to 1000°F, air cooled to 140°F, triple tempered at 1025°F (magnification of radial size change, 1000X)

Fig. 44. Out-of-roundness distortion in M2S (1.0% C) high speed steel bars[40]

Processing of P/M Tool Steels

The data presented in the previous section apply to P/M tool steel that is processed by hot consolidation followed by hot rolling to produce bar stock. The starting powder was inert gas atomized as a prealloy. There are other methods of manufacturing tool steel products from powder. These processes are briefly described below.

HIP plus hot work is the most common method of making P/M tool steel mill stock. Powder is gas atomized, containerized under a partial vacuum, hot isostatically pressed (HIP) to full density and then hot rolled to final mill shape. Intricate shapes can be produced by HIP with no secondary working.

Extrusion Plus Hot Work. Extrusion of containerized material can be performed in place of HIP. Ratios above 9:1 result in complete consolidation. Secondary hot working such as rolling guarantees high integrity of particle/particle bonding.

High-Temperature Sintering. Several versions of high-temperature sintering are practiced[41-44]. These include the Fuldens process and the CAP process.

Universal Cyclops holds the patents on the CAP process[41]. (CAP stands for Consolidation at Atmospheric Pressure.) Gas-atomized powder, low in oxygen level, is coated with a fluxing agent, which is boric acid for steels. The treated powder is charged into Pyrex glass molds and sealed under vacuum. These molds are charged into a furnace and held at a temperature close to but below the liquidus temperature of the alloy. Densification occurs as the fluxing agent accelerates sintering. Upon removal from the furnace, the glass mold is self-stripping, leaving a billet that is between 95 and 100% of theoretical density, depending on the heating conditions. These billets are normally hot worked to mill shape.

The Fuldens process starts with a cold pressable powder. (Water-atomized tool steel powders are available that are cold pressable, or gas-atomized powders can be treated to make them cold pressable.) A compact is cold pressed and then vacuum sintered at a high temperature. A transient liquid phase is formed which accelerates densification during sintering. Products between 95 and 100% of theoretical density are produced. Both complex shapes and simple shapes may be produced. If desired, an additional working step, such as hot isostatic pressing of the bare part, may be used as a supplement to the high-temperature sintering step.

Ferrous Superalloys

The virtues of P/M technology open the field of superalloys to expansion. Ferrous superalloys can benefit from unique chemical compositions, dispersion strengthening techniques such as mechanical alloying, and rapid solidification technology. Iron-aluminum alloys offer densities lower than those of stainless steels and good elevated-temperature oxidation resistance. Such alloys owe their future to P/M processing.

Summary

Ferrous powder metallurgy has been shown to be a vast field in that many alloys and processes are utilized to produce components for a wide range of applications. In each of the applications introduced, P/M technology was utilized because of the advantages it offered. These advantages include economics, shapemaking capability, uniqueness of material or part, part performance, and ability to automate the production process. Ferrous P/M will continue to grow as designers become more educated about P/M and as researchers develop new materials and processes for implementation by the designers and parts manufacturers.

References

1. F. V. Lenel, "Powder Metallurgy–Principles and Applications," MPIF, 1980
2. H. Lubanska, *J. of Metals*, 22, 45, 1970
3. E. Klar and W. Shafer, "Powder Metallurgy for High Performance Applications," edited by J. Burke and V. Weiss, Syracuse University Press, 1972

4. C. F. Dixon, *Canadian Met. Quarterly*, 12, 809, 1973
5. P. Rao, "Shape and Other Properties of Gas Atomized Powder," Ph.D. Thesis, Dept. of Chemical Engineering, Drexel University, Philadelphia, 1973
6. R. J. Grandzol, "Water Atomization of 4620 Steel and Other Metals," Ph.D. Thesis, Dept. of Chemical Engineering, Drexel University, Philadelphia, 1973
7. Powder Metallurgy Equipment Manual, MPIF, 1977
8. J. S. Hirschhorn, "Introduction to Powder Metallurgy," APMI, 1969
9. B. I. Edelson and M. M. Baldwin, *Trans. ASM*, 55, 230, 1962
10. A. Squire, *Trans. AIME,* 171, 485, 1947
11. G. D. McAdam, *J. Iron Steel Inst.,* Vol 168, 1951, p 346
12. Metal Powder Industries Federation Standards
13. K. H. Miska, *Materials Engineering,* Aug. 1976
14. T. A. Harker, Hoeganaes P/M Technical Conference, 1978
15. J. Y. Dress, Hoeganaes P/M Technical Conference, 1978
16. Private Communication with D. Gustafson, Imperial Clevite Inc.
17. C. Durdaller, "Copper Infiltration of Fe-Base P/M Parts," Hoeganaes Technical Bulletin, 1969
18. H. A. Huhn, S. Pillay and H. Chung, "Computer Aided Preform Design for Powder Forging," *Modern Developments in Powder Metallurgy,* Vol 12, edited by H. H. Hausner, H. W. Antes and G. D. Smith, MPIF, Princeton, NJ, 1981, p 643-662
19. S. Pillay and B. L. Ferguson, Contract DAAE-07-08-C-9115 for U.S. Army TACOM, 1980-1982
20. P. W. Lee and H. A. Kuhn, "Fracture in Cold Upset Forging–A Criterion and Model," *Met. Trans.,* Vol 4, Apr 1973, p 969-974
21. C. L. Downey and H. A. Kuhn, "Application of a Forming Limit Concept to the Design of Powder Preforms for Forging," *Journal of Engineering Materials and Technology,* Vol 97, No. 4 of ASME Series H, 1975, p 121-125
22. H. A. Kuhn, "Fundamental Principles of Powder Preform Forging," *Powder Metallurgy for High Performance Applications*, Syracuse University Press, Syracuse, NY, 1972, p 153-169
23. H. A. Kuhn, "Deformation Processing of Sintered Powder Materials," *Powder Metallurgy Processing–New Techniques and Analyses,* edited by H. A. Kuhn and A. Lawley, Academic Press, New York, 1978, p 99-138
24. B. L. Ferguson, "P/M Forging of Components for Army Applications," Tri Service MTAG Program Status Review, Daytona Beach, Sept 1979, p F-1 to F-18

25. B. L. Ferguson, "Toughness and Fatigue of Iron-Base P/M Forgings," Ph.D. Thesis, Drexel University, Philadelphia, 1976
26. S. K. Suh, "Prevention of Defects in Powder Forging," Ph.D. Thesis, Drexel University, Philadelphia, 1976
27. S. K. Suh and H. A. Kuhn, "Three Fracture Modes and Their Prevention in Forming P/M Preforms," *Modern Developments in Powder Metallurgy,* Vol 9, edited by H. H. Hausner and P. V. Taubenblat, MPIF, Princeton, NJ, 1977, p 407-425
28. T. Tabata and S. Masaki, "A Fracture Criterion for Porous Materials and Its Application to the Shape of Sintered Preforms in Forging," *Journal of Engineering Materials and Technology,* Vol 99, No. 1 of ASME Series H, Jan 1977, p 16-22
29. J. P. Cook, "The Effect of Sintering Temperature and Flow on the Properties of Ni-Mo Steel Hot P/M Formed Material," SAE Paper 740982, 1974
30. F. T. Lally, I. J. Toth and J. DiBenedetto, "Forged Metal Powder Products," Final Technical Report SWERR-TR-72-51 on Army Contract No. DAAF01-70-C-0654, Nov 1971
31. H. Ferguson, "The Manufacture of Ball Bearing Races via Cold Forming of P/M Preforms," *Modern Developments in Powder Metallurgy,* Vol 11, edited by H. H. Hausner and P. V. Taubenblat, MPIF, Princeton, NJ, 1977, p 173-180
32. B. L. Ferguson, H. A. Kuhn and A. Lawley, "Fatigue of Iron-Base P/M Forgings," *Modern Developments in Powder Metallurgy*, Vol 9, edited by H. H. Hausner and P. V. Taubenblat, MPIF, Princeton, NJ, 1977, p 51-76
33. B. L. Ferguson, S. K. Suh and A. Lawley, "Impact Behavior of P/M Steel Forgings," *Int. J. of P/M & P/T,* 11(4), 1975, p 263-275
34. A. Lawley, "Analysis of Mechanical Property–Structure Relationships in Powder Forgings," ibid, p 139-171
35. F. Hanjeko, "Mechanical Property Anisotropy of P/M Hot Forged Materials," *Modern Developments in Powder Metallurgy,* Vol 10, edited by H. H. Hausner and P. V. Taubenblat, MPIF, Princeton, NJ, 1977, p 73-88
36. F. T. Lally and I. J. Toth, "Forged Metal Powder Gears," Technical Report No. 11960, Army Contract #7-72-C-0277, Sept 1974
37. K. F. Chesney, "Forging Powder Metal Preforms," Technical Report No. 12151, U.S. Army TACOM, Warren, Michigan, Apr 1974
38. M. M. Jost and C. R. Cook, "Precision Forging Process for P/M Fins," *P/M in Defense Technology,* Vol 5, MPIF, Princeton, NJ, 1980, p 115-128

39. R. M. Szary and R. Pathak, "Sinta-Forge–An Efficient Production Process for High Fatigue Stress Components," P/M Technical Conference sponsored by Hoeganaes Corp., Oct 1978
40. A. Kasak, G. Steven and T. A. Neumeyer, "High-Speed Tool Steels by Particle Metallurgy," SAE Paper 720182, 1972
41. Anon., "Major P/M Expansion at Universal-Cyclops Features New Consolidation Process," *Industrial Heating,* 48(12), 1981, p 8-10
42. M. T. Podob, L. K. Woods, P. Beiss and W. Huppman, "The Mechanism of Sintering High Speed Steel to Full Density," *Modern Developments in Powder Metallurgy*, Vol 13, edited by H. H. Hausner, H. W. Antes and G. D. Smith, MPIF, Princeton, NJ, 1981, p 71-92
43. K. M. Kulkarni, A. Ashurst and M. Svibar, "Role of Additives in Full Dense Sintering of Tool Steels," ibid, p 93-120
44. E. A. Carlson, J. E. Hansen and J. C. Lynn, "Characteristics of Full-Density, P/M Tool Steel and Stainless Steel Parts," ibid, p 121-136
45. M. Goransson, H. G. Larsson and A. Hede, "Breakthrough in Special Steel Production by P/M," ibid, p 307-324
46. Crucible Data Sheet on CPM 10V Tool Steel, Issue #2

Chapter 5

Stainless Steel Powder Metallurgy

Donald L. Dyke
Sintered Specialties Incorporated

Harry D. Ambs
SCM Metal Products
SCM Corporation

Stainless steel powder metallurgy parts represent an important and interesting segment of the P/M industry. The volume of P/M stainless steel parts produced has more than doubled in the last decade. Figure 1 shows that these parts represent only 1.4% of the total P/M industry by tonnage but about 3.2% of the industry's total dollar sales (figures are for North America).

Most applications of these parts are in environments that are only mildly corrosive, and P/M parts have performed very well. However, stainless steel P/M is still a young and emerging segment of the P/M industry. Figure 2 shows the sales of stainless steel powders in the United States for the past 20 years. This curve shows a generally upward trend interrupted by the recessionary years of 1975 and 1980. Stainless steel P/M is less dependent on automotive applications, because it serves many market segments. Figure 3 shows the approximate distribution of stainless steel P/M parts in 1979. With this broad base, stainless steel P/M obviously has many potential avenues of growth.

Powders

All commercial compacting-grade stainless steel powders are produced by water atomization. This manufacturing technique was developed in the early 1950's and opened

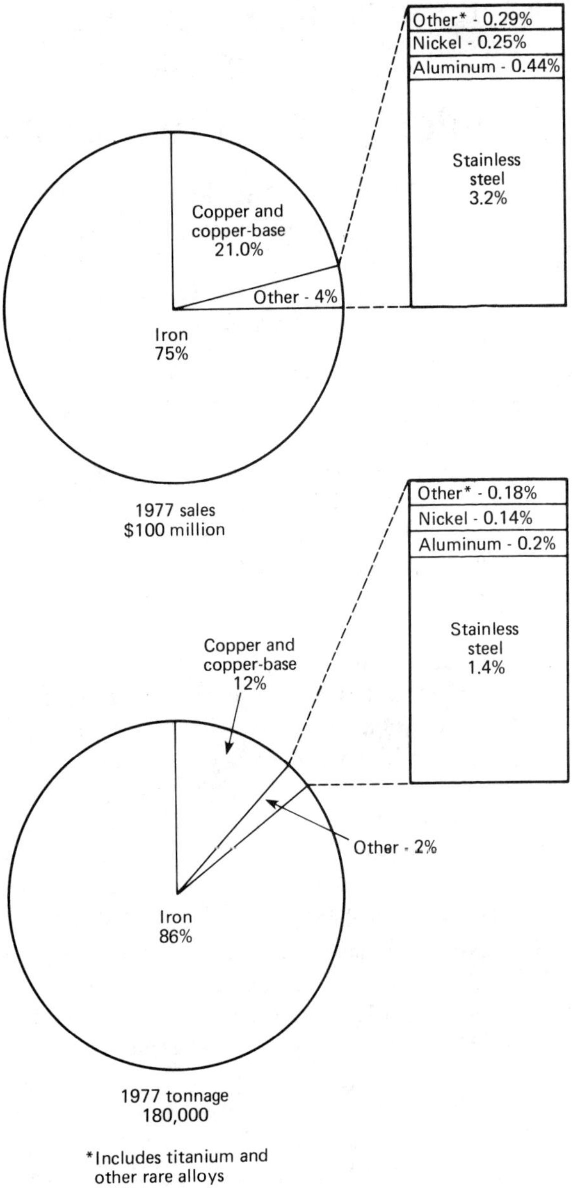

Fig. 1. Distribution of P/M parts made from specialty alloys in 1977

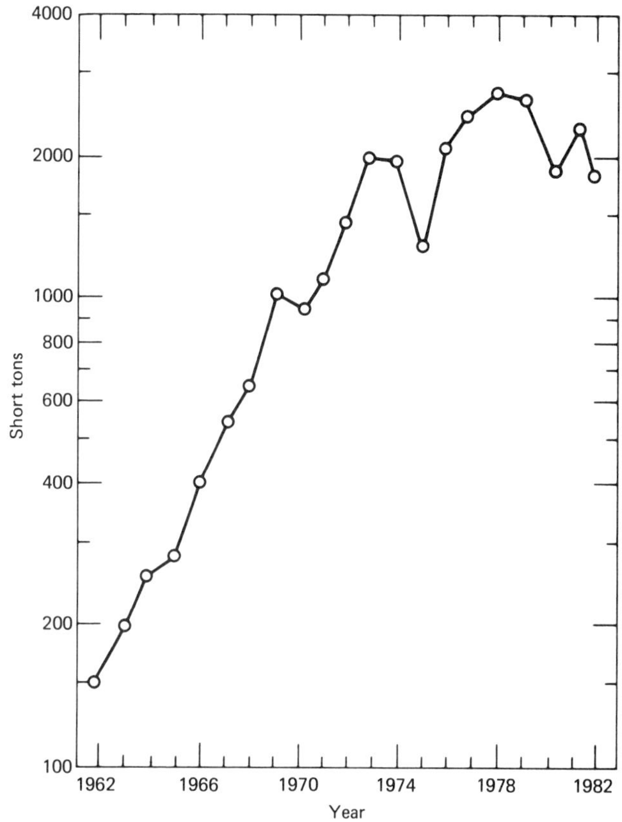

Fig. 2. U.S. sales of P/M stainless steel powders from 1962 to 1982

the door to the growth of stainless steel powder metallurgy. This process produces a powder that is fully alloyed and that has an irregular particle shape and sufficient compressibility and green strength to be suitable for P/M fabrication. A typical sieve analysis of such a powder is shown in Table 1.

The chemical compositions of these alloys (see Table 2) generally meet the AISI limits for wrought alloys, but in practice are held to closer tolerances. Low carbon contents,

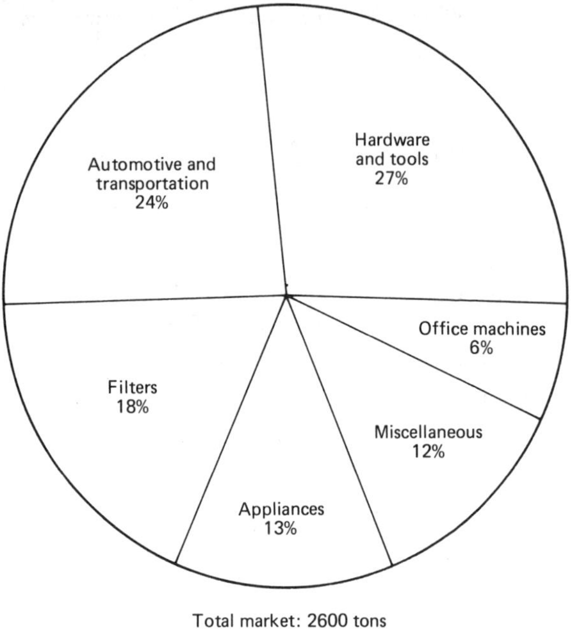

Fig. 3. 1979 U.S. market distribution of P/M stainless steel products

and high nickel contents in the austenitic alloys, contribute to good compressibility.

Powder prices depend on the grade and quantity purchased. Table 3 lists prices of the most popular alloys at the beginning of 1982.

The austenitic grades are the most widely used, with 316L alone accounting for more than 1/3 of all stainless steel powder.

The design and processing of stainless steel P/M parts are subject to the same basic considerations as for other P/M materials. Compared with low-alloy ferrous compositions, however, the stainless parts producer is confronted with higher compacting pressures and lower green strength. Typical compaction behavior is illustrated in Fig. 4. Compacting pressures ranging from 40 to 60 tons/in.^{2} are common in commercial practice.

Table 1. Typical sieve analysis of P/M stainless steel powder

U.S. standard sieve size	Particle size, μm	Percentage (by weight)
+100	>149	0-3
−100 +140	105-149	10-15
−140 +200	74-105	15-25
−200 +325	44-74	20-30
−325	<44	35-50

Table 2. Nominal compositions of stainless steel powders

Grade	Composition(a), %						
	Ni	Cr	Mo	Si	Mn	S	C
Austenitic:							
303L	12.5	17.5	...	0.7	0.35	0.2	0.02
304L	10.5	18.5	...	0.7	0.25	...	0.02
316L	13.0	17.0	2.1	0.7	0.20	...	0.02
Ferritic:							
434L	...	17.0	1.0	0.7	0.25	...	0.02
Martensitic:							
410L	...	13.0	...	0.7	0.40	...	0.02

(a) Remainder Fe.

Table 3. U.S. prices of stainless steel powders in 1982

Grade	Price per pound(a)
303L	\$1.62 to \$1.42
304L	\$1.59 to \$1.39
316L	\$2.11 to \$1.80
410L	\$1.29 to \$0.98
434L	\$1.42 to \$1.10

(a) Commercial quantities, 2000 to 20,000 lb.

For stainless steel compacts, as for compacts of other ferrous powders, green strength is influenced by compacting pressure and type of lubricant (Fig. 5). At best, however, green strength of P/M stainless steel is only about half that for P/M irons, and complex parts require special care in ejection and handling.

As with other powder materials, lubricant choice involves a compromise. Those lubricants which give high green

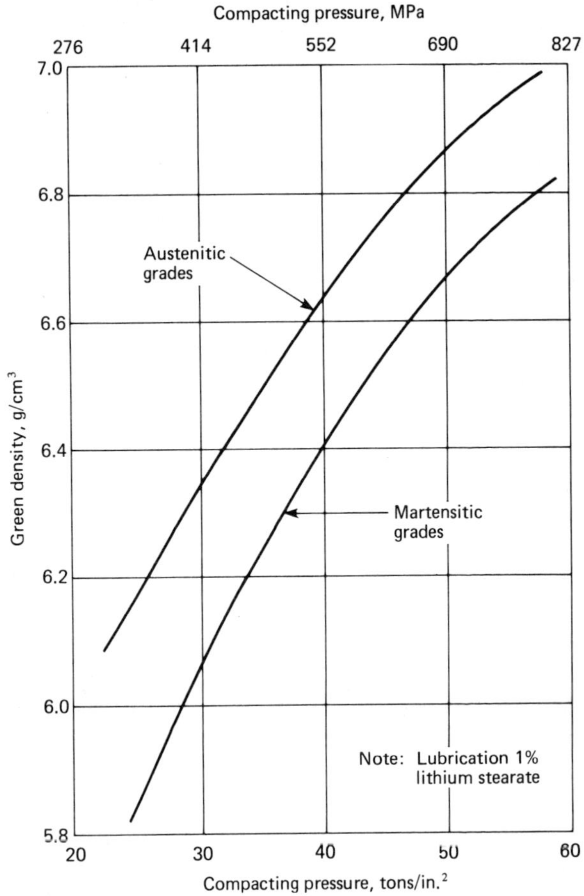

Fig. 4. Typical compaction behavior of stainless steel powders

strength generally cause lower compactibility, and vice versa. Lubricant selection is more important to successful fabrication of stainless steel than to fabrication of any other metal powder. In addition, fabrication of P/M stainless steel parts generally requires the use of carbide tooling, because alloy powders are inherently harder than pure metals. However, the powders available have been devel-

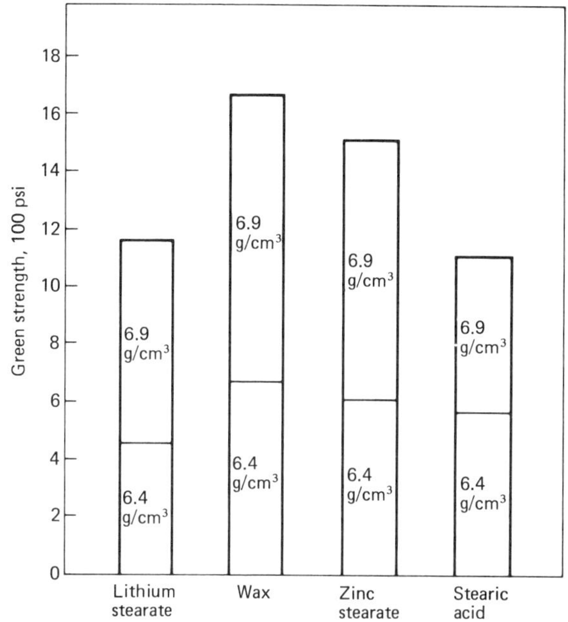

Fig. 5. Effects of lubricant and green density on green strength of type 316L stainless steel compacts

oped and specifically designed for the P/M process and have resulted in the commercialization and growth of the P/M segment.

Sintering

The most important and most critical step in processing stainless steel parts is sintering. During this treatment the lubricant must be removed and the particles bonded together.

Stainless steels for some applications are sintered at 2050 to 2100°F. When improved mechanical properties and corrosion resistance are required, however, temperatures up to 2400°F, and sometimes higher, are used. Continuous mesh belt sintering furnaces are suitable to about 2100°F. At higher temperatures, manual or automatic pusher, walking beam or vacuum furnaces are used.

The most common commercial atmosphere for sintering stainless steel is dissociated ammonia. A dew point of −50 to −60°F at the dissociator is usually adequate for oxide-free work. The oxidation-reduction curve for Cr_2O_3 (Fig. 6) shows that lower dew points are required at lower temperatures if a reducing environment is to be maintained. To ensure that no discoloration occurs during cooling and to allow some latitude in the sintering process, the dissociated ammonia is sometimes dried to a dew point of −80°F or better before being introduced into the furnace.

The nitrogen in the dissociated ammonia is definitely not neutral to stainless steel. The nitrogen content in the material will increase in accordance with the equilibrium diagram shown in Fig. 7. When commercial sintering times and temperatures are used, equilibrium values are reached without difficulty and may range from 0.2 to 0.5 wt %.

Price increases for ammonia and concern regarding availability in recent years have prompted the study of various mixtures of nitrogen and hydrogen gases as protective

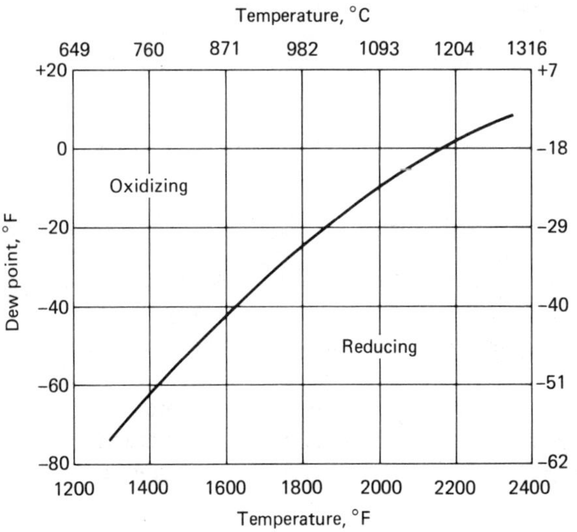

Fig. 6. Oxidation-reduction of Cr_2O_3 with hydrogen

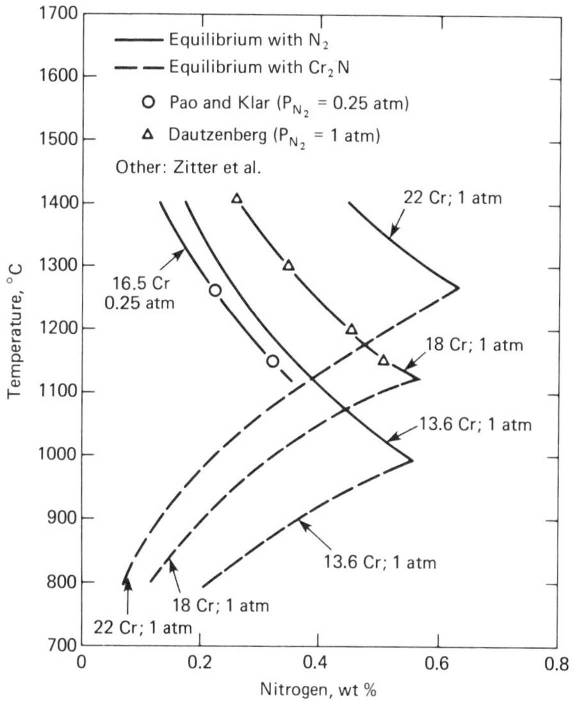

Fig. 7. Nitrogen solubilities of austenitic stainless steels

atmosphere for sintering stainless steel. It has been demonstrated that many parts can be sintered successfully in production with as little as 3% hydrogen in the atmosphere in contrast with the 75% in dissociated ammonia.

The nitrogen content is proportional to the square root of the positive pressure of nitrogen. Thus, sintering in an atmosphere of 90% or more nitrogen results in almost twice the amount of nitrogen as that obtained in dissociated ammonia. Slow cooling results in additional nitrogen pick-up, partly because of increasing nitrogen solubility down to a temperature of about 2000°F and partly because of reduced nitrogen solubility coupled with chromium nitride precipitation below that temperature.

The effects of nitrogen on the mechanical properties of sintered 316L stainless steel can be seen in Fig. 8, 9 and 10 by comparing the properties after sintering in hydrogen and dissociated ammonia. The higher nitrogen contents resulting from high-nitrogen atmospheres appear not to be detrimental in most cases.

Although considerable data on hydrogen sintered stainless steel has been developed, there is very little commercial use of hydrogen because of the high cost. The principal alternative to dissociated ammonia or nitrogen-based atmospheres is vacuum. The acceptance and popularity of vacuum sintering is increasing due largely to energy conservation and ecological considerations. Although conventional cold-wall vacuum furnaces are used, the operation is

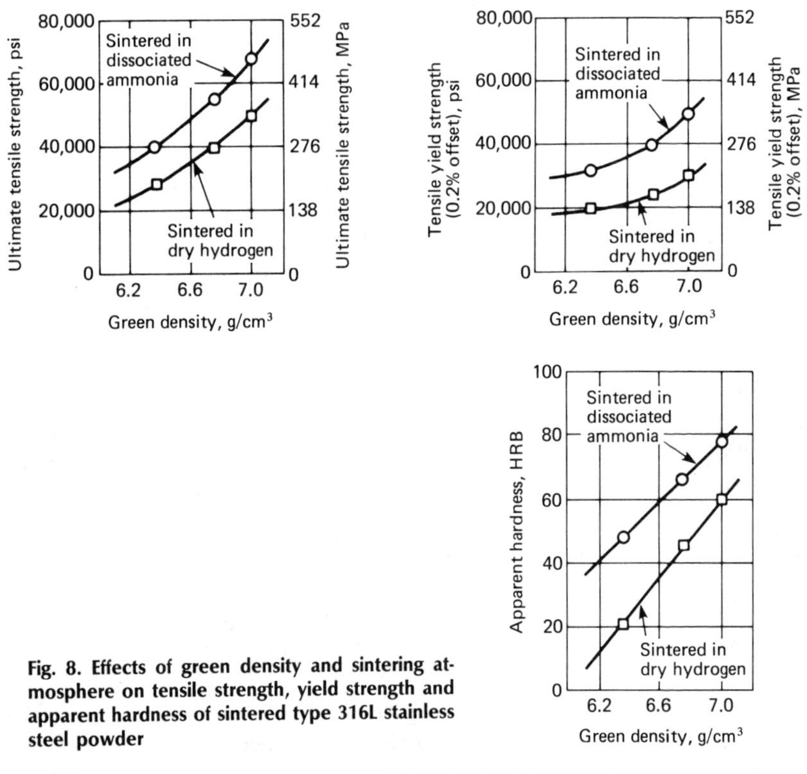

Fig. 8. Effects of green density and sintering atmosphere on tensile strength, yield strength and apparent hardness of sintered type 316L stainless steel powder

more properly described as partial pressure sintering. The vapor pressure of some elements (e.g., chromium) at the sintering temperature is near the pressure that can be achieved in commercial vacuum furnaces equipped only with the mechanical pumping systems. Chromium will evaporate if the furnace pressure falls below its vapor pressure, and corrosion resistance will be seriously reduced. The chromium in a stainless steel can be virtually depleted

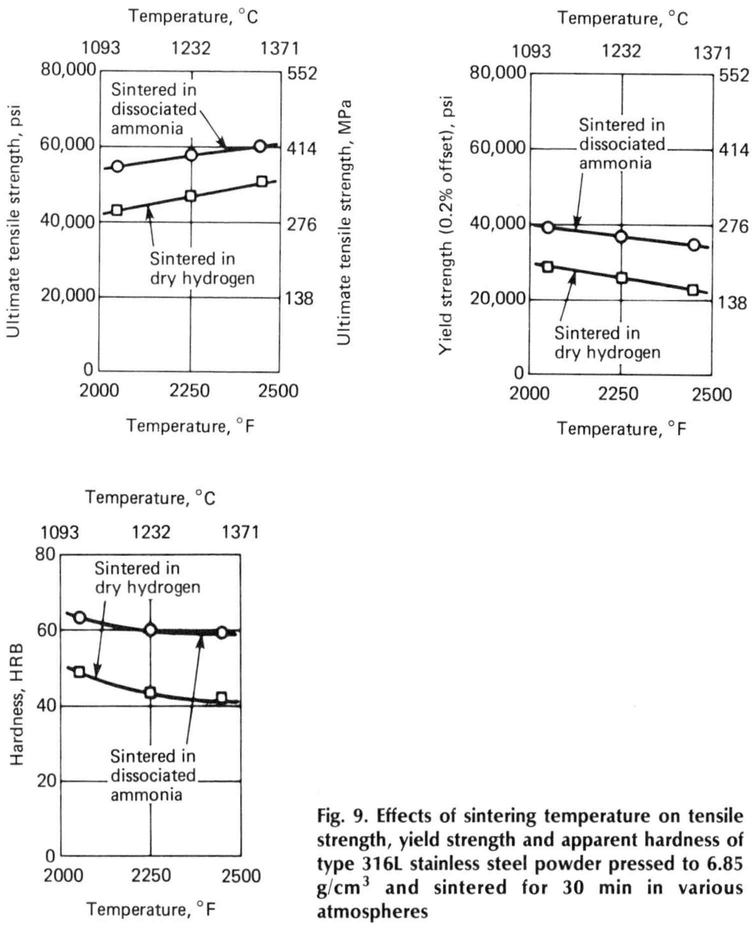

Fig. 9. Effects of sintering temperature on tensile strength, yield strength and apparent hardness of type 316L stainless steel powder pressed to 6.85 g/cm^3 and sintered for 30 min in various atmospheres

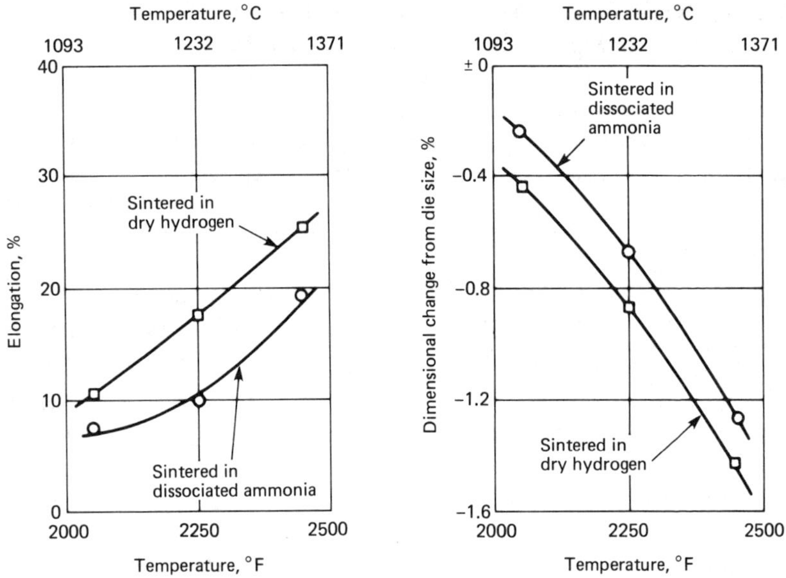

Fig. 10. Effects of sintering temperature on elongation, and dimensional change during sintering, of type 316L stainless steel powder pressed to 6.85 g/cm^3 and sintered for 30 min in various atmospheres

in a typical vacuum sintering cycle if the vacuum level is not properly controlled. The required control is achieved by backfilling the vacuum vessel with a suitable gas to a partial pressure well above the vapor pressure of any of the elements in the alloy. A gas pressure of 200 to 500 μm is typical for sintering of stainless steel at 2400°F. If argon is used as a backfill gas, the mechanical properties will be similar to those obtained in hydrogen. When nitrogen is used for backfilling, the sintered properties will be comparable to those achieved in dissociated ammonia. A circulating gas quench is generally used to provide the desired high cooling rate and the gas composition should be the same as that used for maintaining a partial pressure during heating and soaking.

The properties of sintered stainless steel increase with increasing density. Figure 8 shows the mechanical properties and hardness attainable in 316L stainless steel as a function of green density and sintering atmosphere. Densities of 6.2 to 7.0 g/cm^3 cover a large percentage of parts produced. Higher densities can be achieved with high-temperature sintering or multiple pressing.

The effects of sintering temperature on mechanical properties of 316L are shown in Fig. 9 and 10. Tensile strength increases with increasing temperature whereas yield strength decreases. It is well established that the yield strength of a material decreases as grain size increases. With increasing sintering temperature, the amount of particle bonding (which increases strength) increases, but the rate of grain growth is greater so that the net effect is a slight reduction in yield strength.

It was mentioned earlier that stainless steels exhibit greater shrinkage during sintering than do P/M irons. The dimensional change for 316L is shown in Fig. 10. Shrinkage increases significantly with increasing temperature, thus making dimensional control difficult at high sintering temperatures. Type 434L pressed to 6.8 g/cm^3 and sintered for 30 minutes at 2400°F in dissociated ammonia shrinks 3.5 to 4.0% (0.035-0.040 inches per inch) from die size. Properties of "vacuum" sintered 316L are listed in Table 4, and properties of ferritic 434L are given in Table 5.

The type 410 martensitic grade powder contains less than 0.03% carbon, but sufficient hardening is obtained by sintering in dissociated ammonia to satisfy many applications. When sintering is done in an atmosphere without nitrogen, it is necessary to add approximately 0.15% graphite to the powder. Table 6 shows the effects of sintering atmosphere, graphite additions and heat treat conditions on the strength and hardness of 410 stainless steel at a density of 6.7 g/cm^3 sintered for 30 minutes at 2050°F.

Corrosion Resistance

The primary reason for selecting a stainless steel composition is usually its corrosion resistance. Because the corrosion behavior of P/M materials is not as well documented

Table 4. Properties of type 316L stainless steel sintered for 140 min at 2375°F in a vacuum furnace under 300 μm N_2 and quenched with N_2

Property	Value at compacting pressure of: 40 tsi	50 tsi	60 tsi
Density, g/cm^3	6.9	7.1	7.2
Tensile strength, psi	79,525	85,610	85,940
Yield strength (0.2% offset), psi	51,960	55,650	55,270
Elongation, % in 1 in.	16.0	16.6	18.2

Table 5. Properties of type 434L stainless steel sintered for 30 min at 2400°F in dissociated ammonia

Property	Value at compacting pressure of: 40 tsi	50 tsi	60 tsi
Density, g/cm^3	7.0	7.1	7.2
Tensile strength, psi	58,000	61,600	65,800
Yield strength (0.2% offset), psi	36,700	39,100	41,250
Elongation, % in 1 in.	17	19	21

Table 6. Properties of sintered type 410 stainless steel

Processing treatment	Graphite added, %	Sintering atmosphere	Tempering temperature °F	°C	Tensile strength ksi	MPa	Apparent hardness, HRB
As sintered and cooled in water-jacketed zone of furnace	0	NH_3	...	...	105	724	102
	0.10	NH_3	400	204	99	682.6	103
	0	H_2	...	...	57	393	68
	0.10	H_2	350	177	103	710.2	95
Reheated in NH_3 and oil quenched from 1750°F (950°C)	0	NH_3	400	204	91	627.4	106
	0.10	NH_3	430	221	102	703.3	102
	0	H_2	...	...	109	751.6	106
	0.10	H_2	430	221	104	717.1	105
Reheated in H_2 and oil quenched from 1750°F (950°C)	0	NH_3	400	204	106	730.9	104
	0.10	NH_3	400	204	108	744.7	104
	0	H_2	400	204	93	641.2	95
	0.10	H_2	430	221	116	799.8	101

as that of comparable wrought materials, it is advisable to perform functional environmental tests on P/M parts. The corrosion resistance of sintered stainless steel depends on the processing of the material, on its porosity and on the environment to which it is exposed. Several generalities pertaining to corrosion resistance can be stated:

- High-temperature sintering usually improves corrosion resistance, especially in nitrogen-containing atmospheres.
- The carbon content must be maintained as low as possible. With proper housekeeping and control of furnace atmosphere, it can be held below 0.03%.
- Contamination with iron particles accounts for many instances of development of rust spots in atmospheric exposure or in other relatively mild environments. Isolation of the equipment used to process the stainless steel is a reliable method of solving this problem.
- Corrosion resistance of P/M materials can be improved by modifying chemical composition or by using special surface treatments. Some proprietary materials and treatments are already available and others are being developed.

Recent studies have shown that corrosion resistance increases with increasing part density in acid environments and decreases with increasing density in chloride environments. The reason for the latter phenomenon is that higher density results in smaller pores, which, in turn, are more sensitive to crevice corrosion.

Recent studies also have shown that some of the discrepancies among the effects of sintering atmospheres reported in the literature may be due to differences in oxygen and nitrogen contents. It has been shown, for instance, that the major surface oxide present on atomized powder particles as well as on sintered parts is not chromium oxide but silicon oxide. In nitrogen-containing sintering atmospheres, it is now well established that fast cooling after sintering is essential for preventing additional nitrogen absorption and precipitation of chromium nitrides. This process is analogous to the well-known carbide sensitization which also leads to chromium depletion and loss of corrosion resistance.

Figures 11 and 12 show the microstructures of both acceptable and unacceptable commercially sintered austenitic stainless steel parts.

Prealloyed austenitic stainless steel powders containing small amounts of tin have recently become commercially available. Nitrogen has a low solubility in tin, which is concentrated in the outermost layers of the stainless steel powder particles. Therefore, during sintering in nitrogen-containing atmospheres, the tin acts as a barrier to nitrogen absorption and chromium nitride precipitation, even under conditions of slow cooling. The result is superior corrosion resistance. This effect is most pronounced in lower-density parts, into which nitrogen absorption is very rapid.

Applications

Structural Parts

Stainless steel P/M parts are used in a wide variety of applications in many different industries. The selection of P/M over alternate methods of fabrication is based on the same considerations as for other materials. Like other

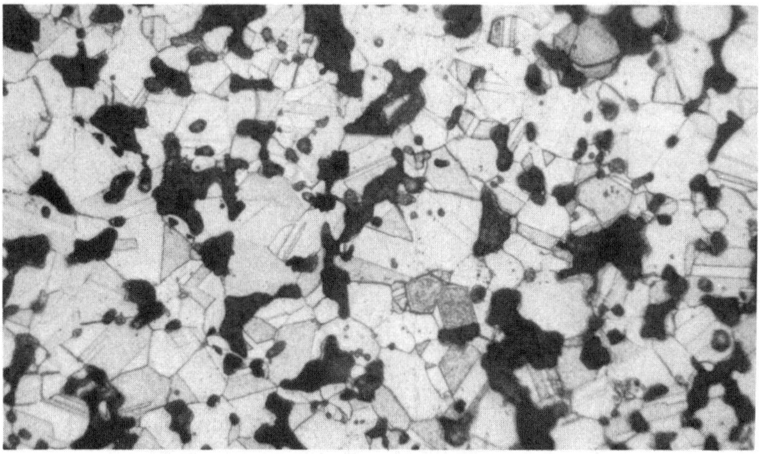

Fig. 11. Microstructure of acceptable commercially sintered type 316L stainless steel, showing twinning bands and some rounding of pores. Magnification, 400×.

Fig. 12. Microstructure of unacceptable commercially sintered type 316L stainless steel, showing heavy chromium carbide precipitation at grain boundaries. Carbon content, 0.12%. Magnification, 400×.

materials, stainless steel parts are machined, impregnated, tumbled and subjected to other secondary operations as required for the application.

Table 7 shows the distribution of parts by market segment.

Figures 13 to 16 show examples of typical P/M stainless steel parts that are found in these various market segments.

Table 7. Breakdown of total volume of P/M stainless steel by market segment for 1973 and 1980

Market segment	Volume, tons 1973	1980
Automotive and transport	820	700
Appliances	330	200
Hardware	120	400
Filters	390	500
Office machines	(a)	100
Miscellaneous	290	100
Total	1950	2000

(a) Not identified in literature in 1973.

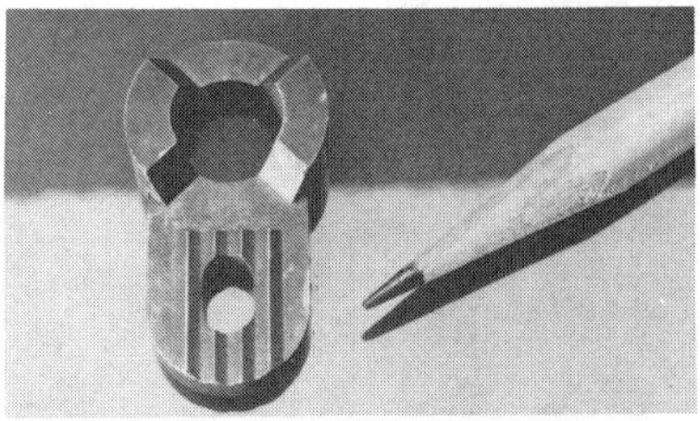

(Left) Type 304L hinge for a refrigerator door. (Right) Type 304L insert for a rubber-molded impeller for a washer pump.

Fig. 13. Typical P/M stainless steel parts for appliances

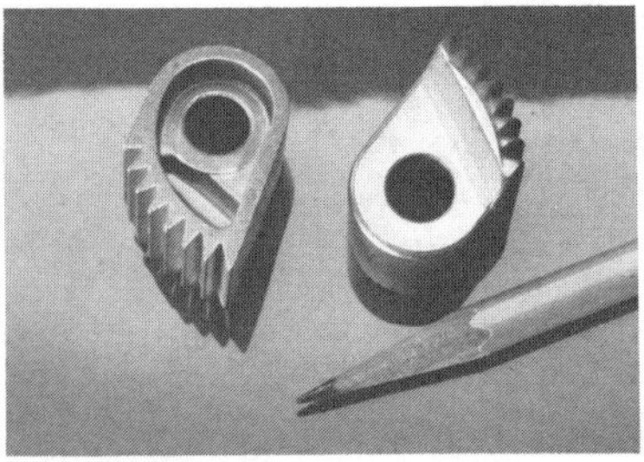

(Left) Type 304L sailboat line cleats. (Right) Various specialty nuts made of type 303L.

Fig. 14. Typical P/M stainless steel hardware parts

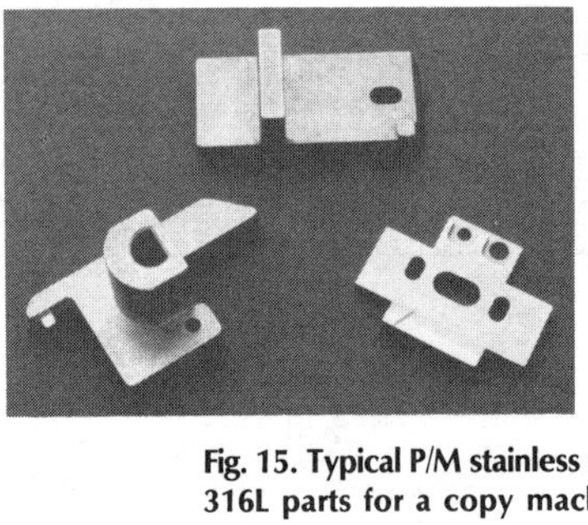
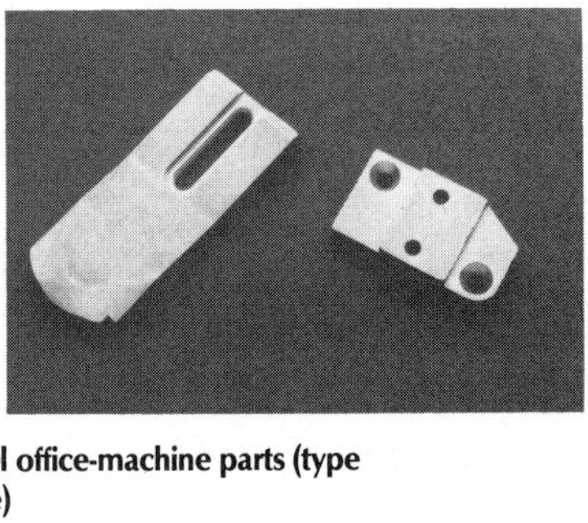

Fig. 15. Typical P/M stainless steel office-machine parts (type 316L parts for a copy machine)

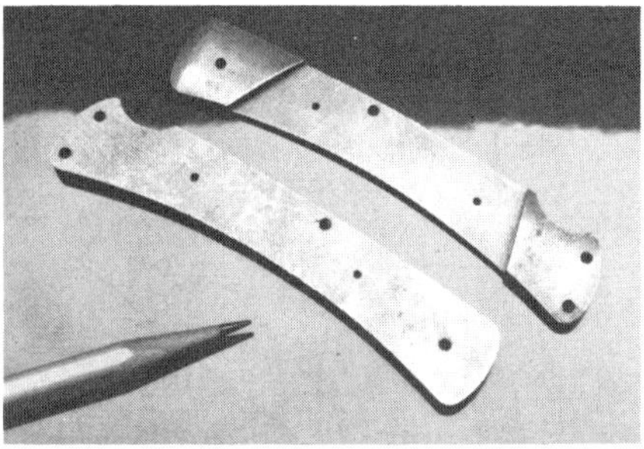

Fig. 16. Typical miscellaneous P/M stainless steel parts (type 316L left and right handle blanks for a hunting knife)

Porous Parts

Porosity is generally an undesirable, although tolerable, characteristic of P/M parts for structural applications. This same characteristic, however, when properly controlled is the key to sintered porous media. Type 316 stainless steel powder is used in the production of many parts with controlled interconnected porosity. Applications include liquid retention as in bearings, filtering, metering of liquids or gases, and sound attenuation in telephones, microphones and hearing aids.

Figure 17 shows the relationship between flow and pressure drop for 1/16-in.-thick type 316 stainless steel with approximate mean pore sizes from 5 to 150 μm.

The effect of porosity of sintered stainless steel on acoustic response is illustrated in Fig. 18. It can be seen that in this specific test circuit attenuation generally increased with increasing density. Even the least resistive sample in the series completely eliminated the peak near 2000 Hz.

Figure 19 reveals that both tensile and yield strength decrease sharply as density decreases (as porosity increases). The strength level of porous P/M parts is adequate, however, for many applications (Fig. 20).

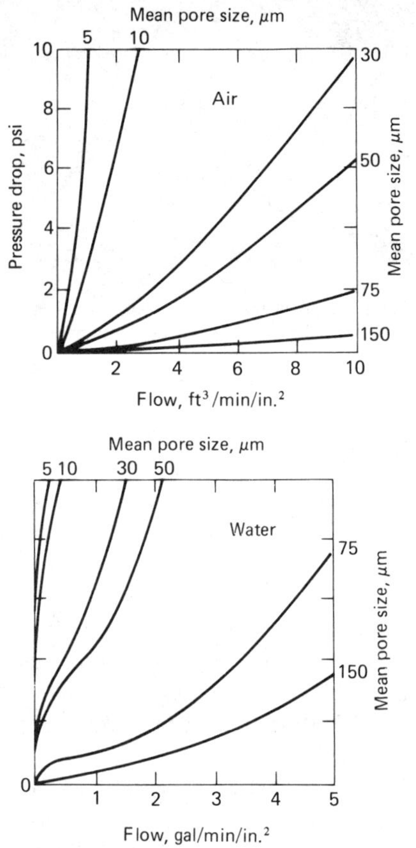

Fig. 17. Relationship between flow and pressure drop for 1/16-in.-thick type 316 porous stainless steel. Mean pore size, 5 to 150 μm.

Future Prospects

We believe that the future of P/M stainless steels lies in the ability to produce improved corrosion-resistant parts by standard processing. Judicious and scientific use of presently available knowledge and know-how alone would

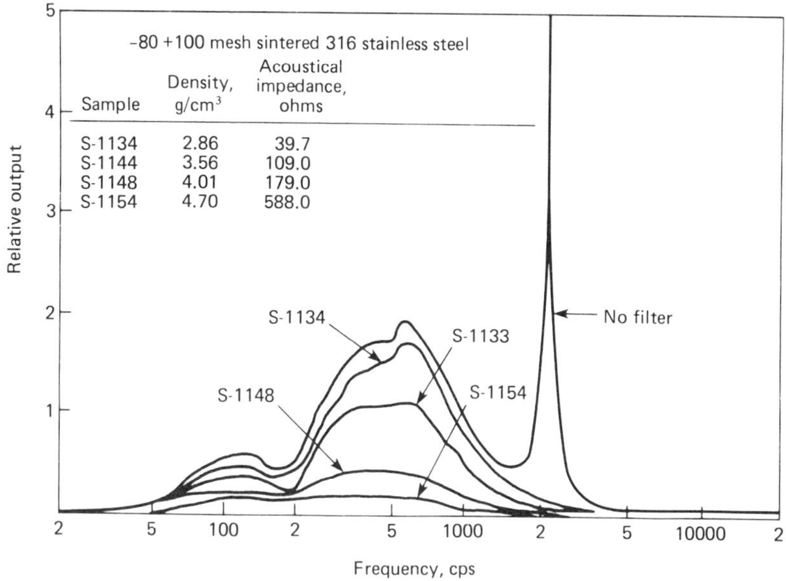

Fig. 18. Effect of porosity on acoustic response of sintered type 316 stainless steel

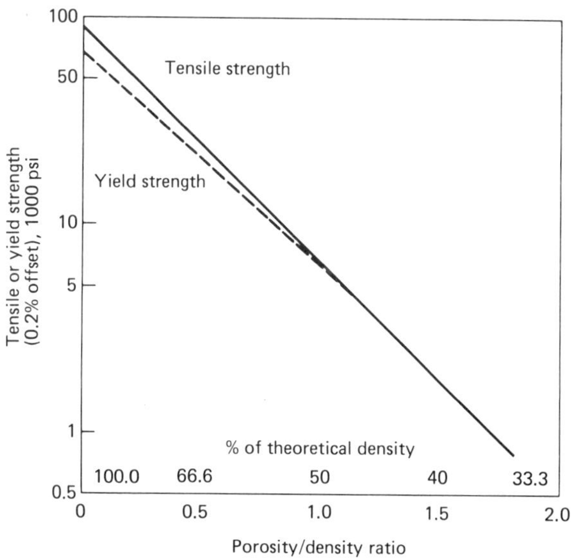

Fig. 19. Effect of density on tensile and yield strengths of P/M stainless steel

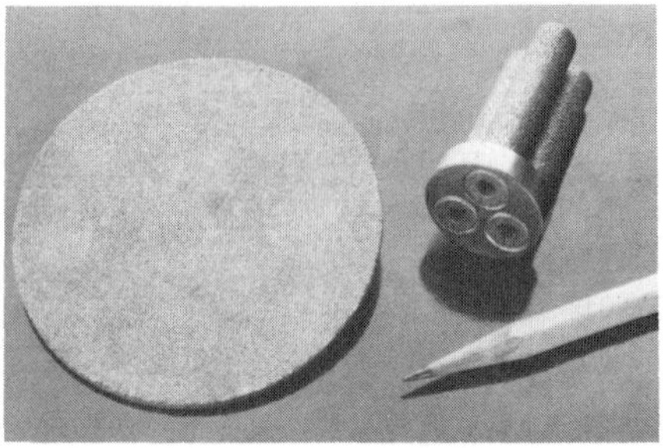

Fig. 20. Typical porous P/M stainless steel filter parts (type 316L plate and three-element filter)

in many cases lead to production of parts possessing adequate corrosion resistance for many applications.

New alloy technology, higher sintering temperatures, and better sintering atmospheres, surface treatments and impregnation treatments, singly or in combination, are expected soon to bridge the remaining gaps between P/M and wrought stainless steels as far as corrosion resistance is concerned. This then would be the basis for a continuation of the growth pattern of P/M stainless steel experienced in the 1960's.

In addition, fully dense stainless steel parts as well as semifabricated shapes made from powder have now become available. The physical and mechanical properties as well as the corrosion resistance of these products are equivalent to those of their wrought counterparts.

Chapter 6

P/M Copper and Copper Alloys

Erhard Klar
Manager, Particle Technology
Glidden Metals Chem./Met. Division
SCM Corporation
Paul E. Matthews
Vice President and Technical Director
U.S. Bronze Powders, Inc.

The use of P/M copper alloys dates back to the 1920's, when porous P/M bronze bearings and bushings became commercialized. Today, these bearings and bushings still account for the major portion of P/M copper and copper alloys. Other important P/M copper-base materials include friction materials, copper-carbon brushes, filters, and structural parts. Total annual consumption of these materials in North America is about 20,000 tons, which is about 1% of the total copper consumption (figures are for 1978). Of all the P/M copper and copper alloy powders used in North America (Fig. 1), about 60% is used for bushings and bearings, 20% in friction applications, 10% as additives to iron powders, 2 to 10% each for copper-carbon brushes and in infiltration, and the remainder in chemical, flake, brazing and other miscellaneous applications.

Powder Production

There are four major methods for making plain copper powders: atomization, electrolysis, oxide reduction, and hydrometallurgy. Brief descriptions of these four processes follow, and typical differences in particle shape and other properties are shown in Table 1 and in Fig. 2 to 6.

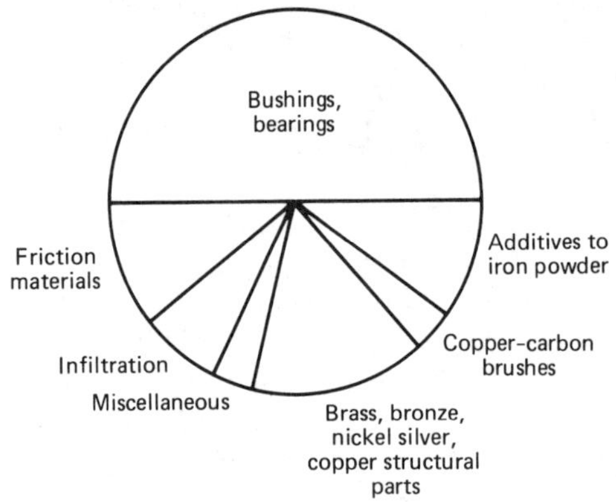

Fig. 1. Relative amounts of P/M copper and copper alloys used for various applications in North America in 1978

Table 1. Characteristics of commercial copper powders

Type of powder	Composition, % Copper	Oxygen	Acid insolubles	Particle shape	Surface area
Electrolytic	99.1-99.8	0.1-0.8	0.03 max	Dendritic	Medium to high
Oxide-reduced	99.3-99.6	0.2-0.6	0.03-0.1	Irregular; porous	Medium
Atomized	99.3-99.7	0.1-0.6	0.01-0.03	Irregular to spherical; solid	Low
Hydrometallurgical	97-99.5	0.2-0.8	0.03-0.8	Irregular agglomerates	Very high

Atomization

The disintegration of a liquid stream by means of an impinging jet of liquid or gas is known as atomization or, more specifically, twin-fluid atomization. This process is used extensively for production of many metal and alloy powders. It is less widely used for production of plain copper powders because of the lower compacting properties of such powders. Figure 7 is a schematic representation of

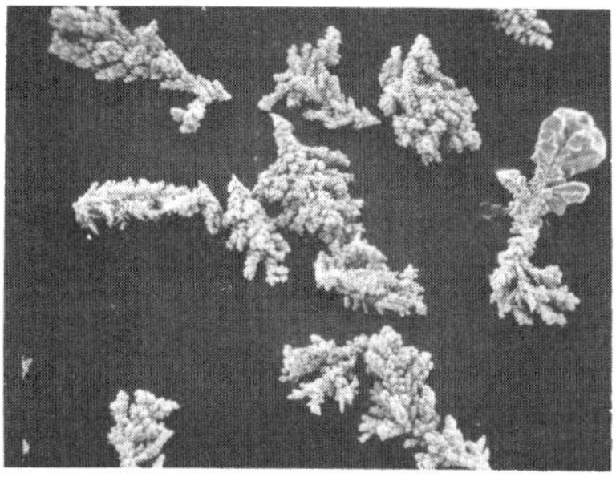

Fig. 2. Electrolytic copper powder (200/325 mesh)

Fig. 3. Oxide-reduced copper powder (–100 mesh)

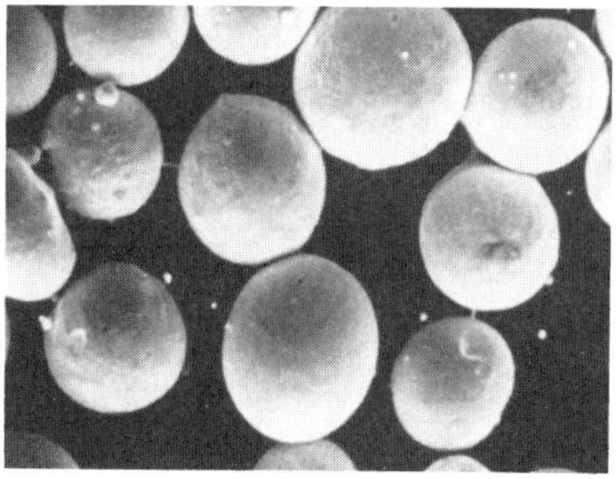

Fig. 4. Gas-atomized copper powder (–100 mesh)

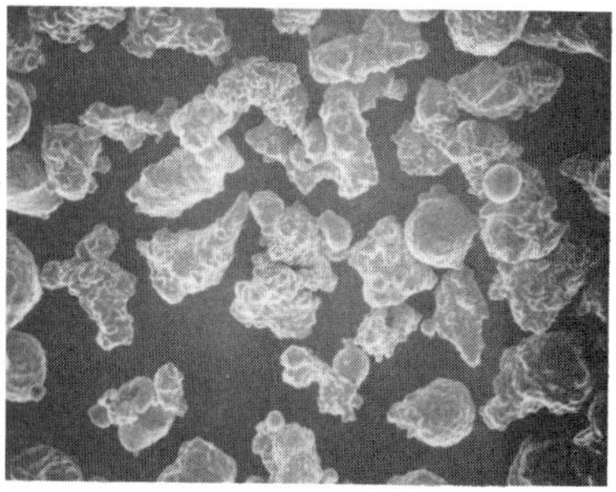

Fig. 5. Water-atomized copper powder (–100 mesh)

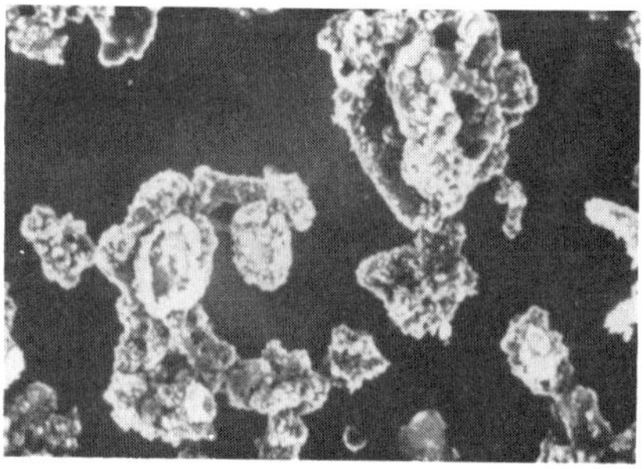

Fig. 6. Hydrometallurgically produced copper powder (–100 mesh)

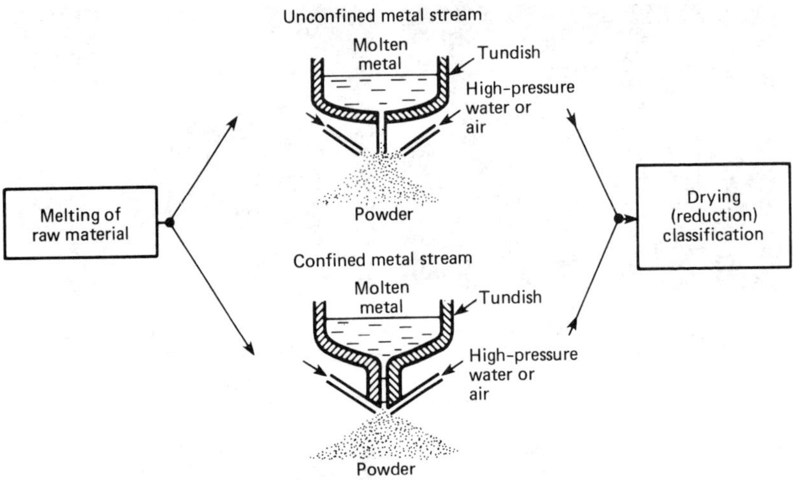

Fig. 7. Processing stages in production of copper by atomization

the processing stages involved in atomization. The atomized powder may be directly dried, classified, and used as such, or it may be subjected to thermal treatment before being used.

Some commercial copper powders classified as atomized powders are sometimes partly oxidized during and subsequent to atomization, and then reduced. This generates partly porous powder particles with improved compacting characteristics.

The particle size of atomized copper powders can be greatly varied by adjusting processing conditions. Increases in gas or water pressure result in smaller particle sizes. Average particle sizes less than 325 mesh (45 μm) are feasible.

Particles of gas-atomized copper powder are more or less spherical, whereas water-atomized copper particles are irregular in shape. The irregularity can be further increased through addition of small amounts of magnesium, calcium, lithium or zirconium to the molten copper prior to atomization.

Of the two atomization designs shown in Figure 7, the "unconfined" (or "free fall") design is used more widely than the "confined" design. The latter design is more difficult to operate but provides improved efficiency and more uniform particle-size distribution.

Oxygen pick-up during atomization of copper is about 0.5 to 0.8% when air is used as the atomizing medium and 0.1 to 0.3% when water is used.

Electrolysis

In production of powder by electrolytic deposition, the processing variables are adjusted so that either a spongy or a brittle polycrystalline deposit is formed rather than a smooth deposit as in electroplating. Many metals and alloys can be deposited in this way. Two examples of commercial significance are copper powder and iron powder. For both, aqueous electrolytes and soluble anodes are used. The low metal overvoltage of copper permits its economic deposition as sponge. Processing stages are shown in Fig. 8. Electrolytic copper powder normally is of high purity; however, impurities more noble than copper are codeposited.

Typical processing conditions include an electrolyte concentration of 5 to 8 g/l of copper and 100 to 160 g/l of sulfuric acid, a bath temperature of about 50°C, a current density of 50 to 100 amp/ft^2 and a cell voltage

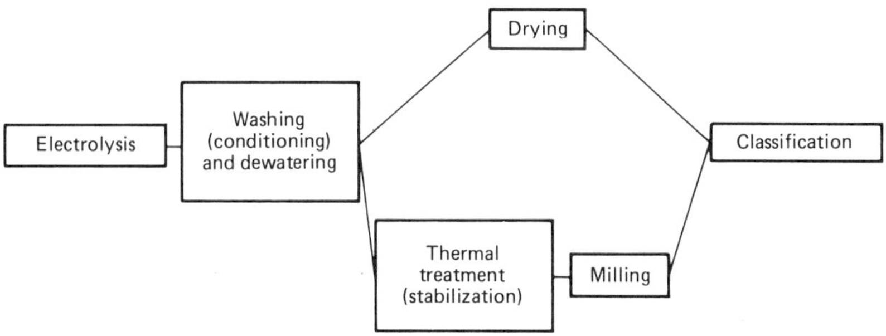

Fig. 8. Processing stages in production of copper powder by electrolysis

of about 1 V. A broad range of particle sizes is possible through control of processing conditions.

The particle shape of as-deposited powder is dendritic or fernlike. A final furnace treatment can alter this shape to such an extent that the processing history of the powder is difficult to determine.

Oxide Reduction

In this process particulate copper oxide is reduced with solid or gaseous reducing agents at elevated temperatures. The product is a more or less sintered porous cake of copper which is ground into powder. Sources of raw material include particulate copper scrap, electrolytic copper, and atomized copper, or copper scale and copper precipitates. The former ones are first roasted to form copper oxide. Selection of raw material is based on purity requirements and end use.

Particle size is controlled through grinding of the starting oxide and the reduced sinter cake. Compacting and other properties may be varied through control of the pore characteristics of the spongy particles.

Hydrometallurgy

A number of metal powders, including copper, cobalt and nickel, can be produced by hydrometallurgical methods. Improvements in the purity of the final powders have made it possible to apply hydrometallurgical methods to the production of copper powders suitable for P/M applications.

The basic processing steps consist of preparing pregnant liquor by leaching ore or another suitable raw material followed by precipitation of the metal from its solution. In the case of copper, the most important precipitation methods are cementation, reduction with hydrogen, and electrolysis.

Only cementation of leached copper ores is used presently on a commercial scale. In this process, the copper-bearing solution is passed over scrap iron, which results in precipitation of copper according to $Fe + CuSO_4 \rightarrow Cu + FeSO_4$. Subsequent separation, washing, thermal reduction, and pulverizing usually produce a copper powder that contains considerable amounts of iron and acid insolubles such as alumina and silica. Contamination with gangue varies and depends on the nature of the pregnant liquor. Low purity in general, and high iron content in particular, restrict the use of cement copper in P/M applications. Its high specific surface area, however, imparts good green strength and makes it useful in friction applications.

The copper powders produced by the various processes may often be used interchangeably, but for some applications certain powder characteristics are very important and result in the preferred use of a specific powder.

Table 2 lists some other copper alloy powders and nonmetallic materials which make up the raw materials for such powder metallurgy products as bronze bearings, friction materials, structural components, and infiltrated and electrical parts.

Some of the more important applications of P/M copper and copper alloys are summarized in Table 3.

Applications

Self-Lubricating Porous Bronze Bearings

Self-lubricating or permanently lubricated porous P/M bearings and bushings are two of the best examples that

Table 2. Raw materials for P/M copper alloys

Copper
Bronze (90Cu-10Sn; 89.5Cu-10Sn-0.5C)
Brass (90Cu-10Zn; 70Cu-30Zn; 88.5Cu-1.5Pb-10Zn; 68.5Cu-1.5Pb-30Zn)
Cupro-nickel (75Cu-25Ni; 90Cu-10Ni)
Nickel silver (64Cu-18Ni-18Zn; 64Cu-18Ni-16.5Zn-1.5Pb)
Graphite
Oxides
Lubricants

Table 3. Applications of P/M copper and copper alloys

Porous bronze bearings
Friction materials
Structural parts (steels; infiltrated steels; brass; bronze; nickel silver)
Filters
Electrical contacts
High-electrical-conductivity copper parts

can be used to illustrate the uniqueness of the P/M process. Sacrifices in some mechanical properties result from the use of a less than fully dense part, but the fact that the pores can be used as oil reservoirs more than compensates for this loss. During use, as the bearing gets hot, the oil expands and flows to the bearing surface. On cooling it is drawn back into the pores by capillary action. These bearings are made by pressing elemental powder blends of copper and tin, followed by sintering at relatively low temperatures (1500 to 1600°F) for short periods of time (15 min is typical). Microstructural changes that occur during this relatively short sintering time, from formation of the tin liquid phase to a homogeneous bronze alpha phase, are shown in Fig. 9 (top) and 9 (bottom) respectively.

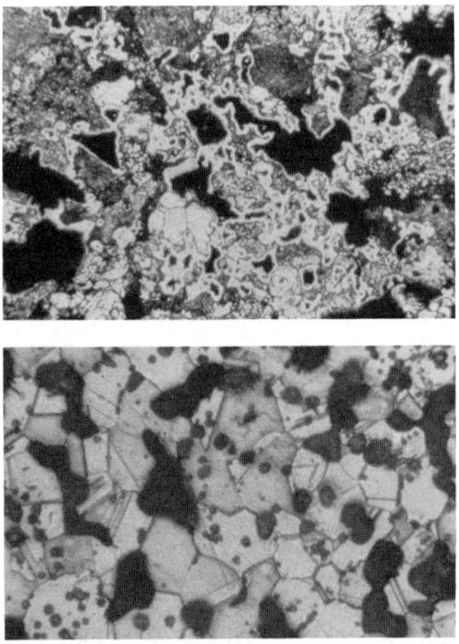

Note unalloyed tin after 1 min of liquid-phase sintering and complete alloying to α-bronze after 10 min. Large black areas are interconnecting pores which provide reservoirs for lubricant.

Fig. 9. Microstructures of P/M 90Cu-10Sn bronze parts after 1 min (top) and 10 min (bottom) in hot zone

An assortment of such bearings is shown in Fig. 10. The most common shapes are simple or flanged bushings, but self-aligning bearings with spherical external surfaces are also possible. For dimensional accuracy, the parts are coined or re-pressed after sintering. Force fitting into a housing is easily accomplished (Fig. 11). These bearings are widely used in the automotive industry, in home appliances, in the leisure industry, and in industrial equipment (Table 4). They run quietly and may be used in vertical positions for which solid bearings would normally be impractical because of lubricant runout. They are

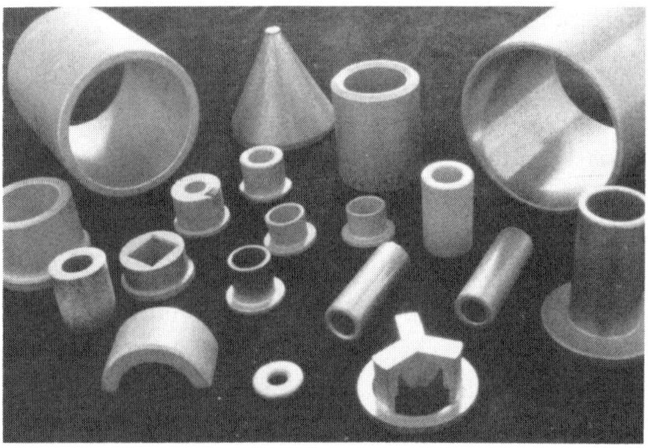

Fig. 10. Assorted P/M bronze bearings

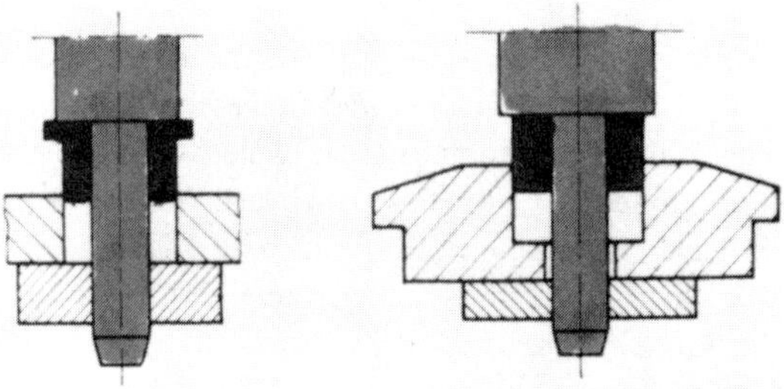

Fig. 11. Force fitting of P/M bronze bearing into housing (Ref. 5)

Table 4. Applications of self-lubricating sintered bronze bearings (fractional horsepower electric motors)

Automotive components:	*Leisure industry; business machines:*
Starters	Phonographs
Light generators	Record changers
Oil and water pumps	Tape recorders
Windshield wipers	Business machines
Hood and window raisers	Electric clocks
Home appliances:	*Industrial equipment:*
Dishwashers	Textile machines
Clothes dryers	Packaging machines
Washing machines	Electric fans
Sewing machines	
Vacuum cleaners	
Refrigerators	
Food mixers	

particularly useful if it is difficult to lubricate the part, such as in a refrigerator motor or where oil splashing may interfere with the operation of the machine.

Typical compositions are Cu-10Sn, Cu-10Sn-0.5C, Fe-2.7Cu-0.8C and Fe-10Cu-0.2C. Pore volumes range from 25 to 35%. These bearings are marketed either dry or saturated with oil. Their sizes range from about 1/32 to about 3 in. in diameter.

The properties that are critical to the performance of a porous bearing are those that are related to: (*a*) maximum oil retention, (*b*) interconnection and isotropy of porosity, (*c*) mechanical strength and (*d*) close dimensional tolerances.

A number of industry standards are available. An example of an MPIF standard is presented in Table 5. This standard provides values of composition, density range, amount of interconnected porosity, ultimate tensile strength, compressive strength, elongation, and the so-called K-factor, which is a measure of radial crushing strength. In addition to these properties, standards exist on dimensional tolerances, running clearances, and guide values for the PV factor—i.e., for permissible loads as a function of shaft velocity. The life of a bearing is closely related to its PV factor (Fig. 12).

Some of the advantages and limitations of porous bronze bearings are summarized in Table 6. The self-lubricating feature simplifies machine design. Due to mass production, these bearings often are less expensive than bearings machined from castings, but they are lower in mechanical strength due to the presence of porosity. Porous metal bearings are most popular in applications involving high velocities and light loads. They are less useful under conditions of impact or fatigue loading. The porosity also reduces thermal conductivity, so that allowable maximum frictional losses are lower than those for nonporous bearings.

Shapes of P/M bronze bearings are limited to those which can be formed by axial compression. There are limits regarding maximum length-to-diameter ratio. Also, there are limits on length-to-wall-thickness ratio because of the requirement of reasonable uniform porosity along the axis of compression. Finally, compositions are limited to alloys amenable to the P/M technique.

P/M techniques are used in the production of nonporous, fully dense bearings such as the so-called dry metal graphite bearings (used under poor lubrication conditions and at intermediate temperatures of 250 to 350°C). These materials are based on brass and bronze (also on Fe and Ni) and contain about 10 to 50 vol % graphite. They are made mainly by hot pressing and can operate under adverse conditions of dirtiness and dustiness (applications: hot water pumps, coal cutting machinery, textile industry). Other nonporous fluid-lubricated bearings made by nonconventional powder metallurgy techniques are based on copper-lead alloys, copper-tin-lead alloys* (often as steel-backed bearings), and lead- or tin-base white (Babbitt) metals. They offer some technical advantages over bearings made by casting. Depending on specific composition and structure, these bearings are recommended for various combinations of load, velocity, fatigue and corrosion resistance (passenger-car engines, compressors, gear boxes, etc.).

---

*Typical compositions: Cu-25Pb-0.5Sn (prealloyed); Cu-25Pb-3.5Sn (prealloyed); Cu-10Pb-10Sn (prealloyed); Cu-48Pb-1.5Sn (blend).

Table 5. MPIF standard for P/M bronze

P/M MATERIALS STANDARD **APPROVED: 1961** **REVISED: 1969**

Material	Designation
P/M Bronze	CT-0010-R

Line		Element	Cu	Sn	C	Fe				
1	Chemical Composition (per cent)	From	86.3	9.5	0	0				
		To	90.5	10.5	1.7 (1)	1.0				
		Other Elements	Total by difference equals 0.5 max.							

Line	Col. 1		2	3	4
2	Condition		As-Sintered	Heat Treated	
3	Density	g/cc	6.4 – 6.8		
4	Allowable Density Variation	g/cc			
5	Interconnected Porosity	%	18 min.		
6	Properties, Mechanical		Note: Typical properties based on reference tests (see Properties page 4, Standard 35)		
7	Utimate Tensile Strength	psi	14,000		
8	Tensile Yield Strength 0.2% Offset	psi			
9	Compressive Yield Strength 0.1% Offset	psi	11,000		
10	Elongation	%	1.0		
11	Apparent Hardness				
12	"K" Strength Constant		26,500		
13	Impact Strength (unnotched) *Charpy; †Izod	ft.-lbs.			
14	Fatigue Strength	psi			
15	Special Properties				
16	Typical Applications		Bearings		

Table 5 continued

Col.	1		2	3	4
17	Properties (Continued)				
18	Young's Modulus	psi			
19	Poissons Ratio				
20					
21					
22					
23					
24					
25					
26					
27	Comparable Specifications				
		ASTM	B 438 Grade 1, Type II		
		SAE	841		
		Military	Mil-B-5687-C Type I, Comp. A (2)		
		Other			
28	Comments: (1) Commonly graphite. A maximum of 1.7% of another type solid lubricant may be substituted when authorized by the purchaser. (2) Mil-B-5687-B Type 1, Composition A does not allow 1% max. iron. *Upon metallographic examination, the microstructure of the material shall be substantially alpha bronze and contain no free tin visible at 300X.*				

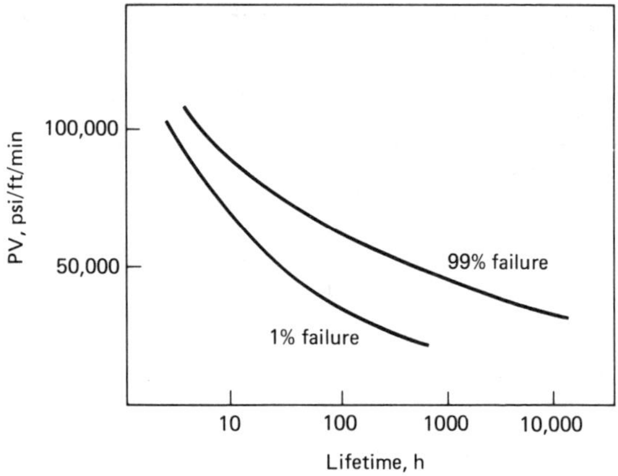

Fig. 12. PV factor vs life for self-lubricating P/M bronze bearing (Ref. 6)

Table 6. Advantages and limitations of porous bronze bearings

Advantages:
Self-lubricating
Minimum oil drip
Use in inaccessible positions
Inexpensive
Limitations:
Impact and fatigue loading
Frictional losses
Size of bearing
Shape
Composition

Friction Materials

Sintered metal friction materials were developed in the 1920's and commercialized in the early 1930's by Wellman. They are used in applications involving the transmission of motion through friction (clutches) and for deceleration and braking. Today, P/M friction materials account for about 15% of the total friction materials industry. Competitive materials include cork, steels, resilient fibrous compounds, resin- and rubber-containing compounds, and asbestos materials. Metallic fillers also are used in combination with asbestos and rubber. Metal-base friction materials are stronger and more heat-resistant and were developed in response to energy inputs and temperatures which exceeded the capabilities of the organic-base friction materials used in the 20's and 30's. World War II, with its demands for large quantities of heavy-duty friction materials in military vehicles and aircraft, contributed much to the growth of that industry. More recently, growth has followed the natural growth of the economy and the need to transmit more and more energy through clutches and brakes of the same size. Energy capacity of oil clutches, for instance, was doubled between 1960 and 1970.

Sintered metal friction materials applications may be classified in terms of dry/wet and mild/moderate/severe applications, as shown in Table 7. The majority of the clutch applications are for wet (oil) operation. Figures 13 to 15 show friction elements used as brake linings and clutch facings. In oil applications the coefficient of friction is lower but part life is longer. Also, there are big differences in finishing operations depending on whether parts must perform dry or in oil. Parts operating in oil have grooves on their surfaces. The grooves help to remove the oil from the interface, which raises the coefficient of friction.

Early P/M friction materials were solely copper-base materials. Today, copper-base materials are still being used in all applications, but iron-base compounds have been developed for moderate-to-severe-duty dry applications. Some typical friction material compositions are shown in Table 8 for both dry and wet applications.

Copper-base materials are used mainly where semifluid friction occurs. For dry friction they are suitable only where operating conditions are relatively mild ($<350^\circ C$).

Typical processing techniques for P/M friction materials include hot pressing and sintering at high temperatures for long times, sometimes followed by re-pressing. Fine and reactive powders are preferred. The friction segments usually are brazed, welded, riveted or mechanically fastened

Table 7. Applications of sintered metal friction materials

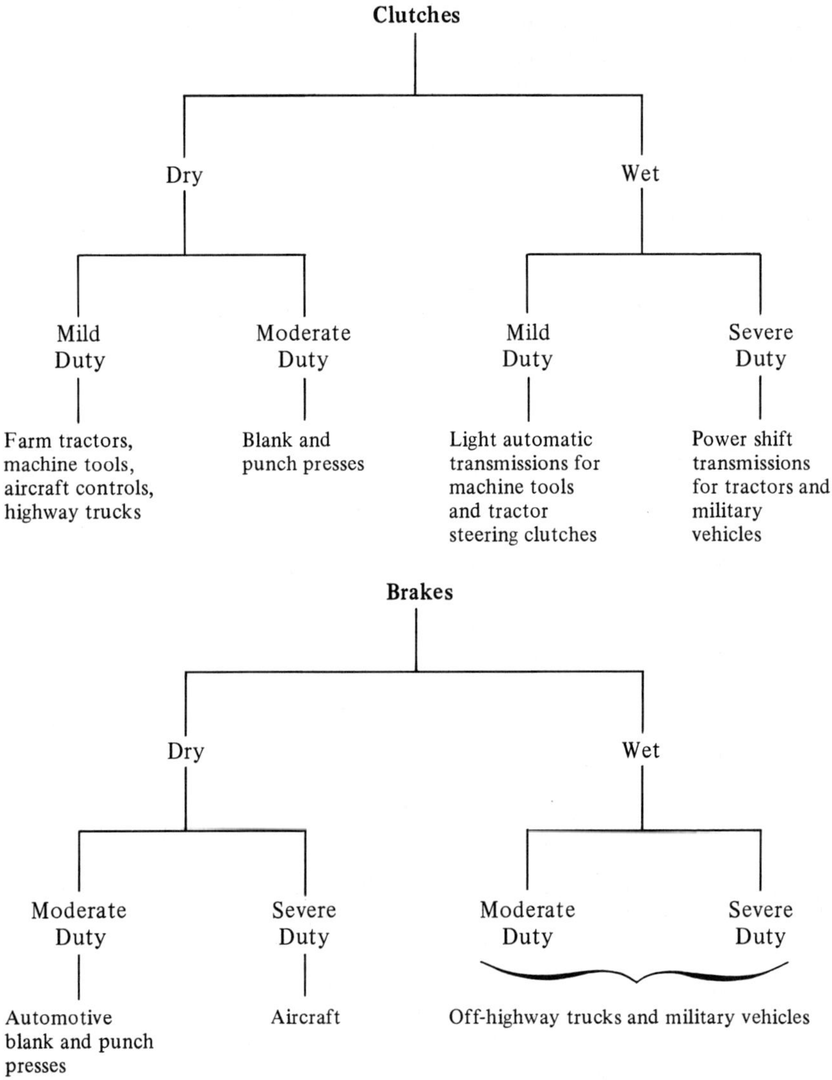

to supporting steel members or are pressure bonded directly to the assembly.

These P/M friction materials are multiphase composites. They contain typically between 4 and 8 ingredients, which

Fig. 13. Copper-base P/M friction elements

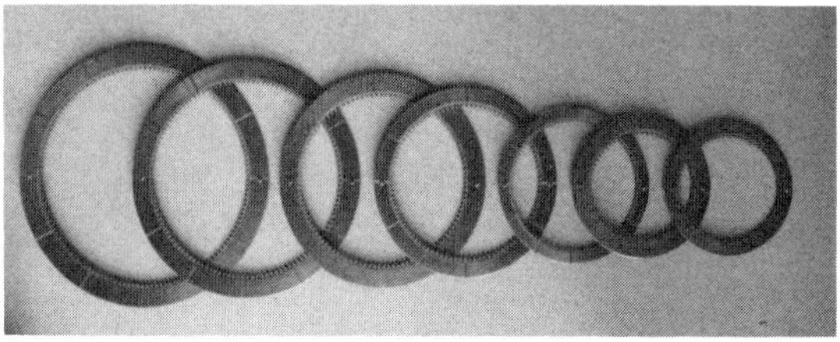

Fig. 14. Copper-base P/M clutch plates (11 to 20-in. OD) used in power-shift transmissions for Caterpillar tractors. (Courtesy of Raybestos)

Fig. 15. Composite copper-base P/M friction pad

leads to innumerable combinations and effects. The functional requirements are very diverse (Table 9). During application, a braking system must exhibit high frictional forces and must be capable of absorbing large amounts of energy. High stresses and much heat are produced. Selection of P/M friction compositions is based on some four or five functional characteristics, which are summarized in Table 10.

Table 8. Typical compositions of P/M friction materials

For dry clutches and brakes:
75Cu-6Pb-7Sn-5C-4MoS_2-3Feldspar
56Fe-14Zn-18C-8Chromite-4SiC
For oil clutches and brakes:
74Cu-3.5Sn-2Sb-16C-4.5Galena

Table 9. Critical performance characteristics of friction materials

Dynamic coefficient of friction
Static coefficient of friction
Static to dynamic coefficients ratio
Durability
Energy capacity
Engagement characteristics
Cost
Wear of opposing member
Fabricability
Temperature coefficient of friction
Time coefficient of friction

Table 10. P/M friction material components used for various functions

Function	Components
Friction, strength, heat conductivity	Matrix/binder: Cu- or Fe-base (Sn, Zn, Pb additions)
Lubrication (seizure prevention; stability)	Dispersed lubricants: graphite, MoS_2, Pb
Abrasion/friction	Abrasive (frictional) components: SiO_2, mullite, Al_2O_3, Si_3N_4
Wear resistance	Cementite, cast iron grit, spinels
Filler	Carbon, minerals

The matrix or binder, which is usually an iron-base or copper-base material, accounts for about 50 to 80% of total weight (>40 vol %). About 5 to 15% consists of a low-melting-point metal such as tin or zinc that alloys with the major constituent through liquid-phase sintering. This provides strength, friction and thermal conductance.

For maximum friction, soft metals with high coefficients of friction are preferred. To avoid gross seizure between friction liner and pad, lubricants are added.

While lubricants (5 to 25 wt %) prevent gross seizure, they do not prevent local welding and metal transfer. To minimize these, up to 20% of an abrasive (often called the frictional component) is added. Since these abrasive components also produce wear, the amount added depends on how much wear can be tolerated in a specific application.

An important requirement is thermal stability, which means that the coefficient of friction and the wear rate do not appreciably change up to a specific temperature. Maximum temperatures are around 350°C for copper-base friction materials and range from 600 to 1100°C for iron-base materials. The wear-resistant components account for up to 10 wt %, essentially for dry applications. Some of these components, such as spinels and mixed metal oxide solutions, may be formed during sintering.

Finally, fillers are used, in amounts up to 15 wt %, to decrease cost.

The coefficient of friction is dependent not only on speed, pressure and temperature of operation but also on composition and powder characteristics of the components. Because of this complexity, optimum compositions are still derived empirically.

P/M Copper-Base Structural Parts

Applications of copper-base P/M materials which rely mainly on the load-bearing capacities of the sintered parts are commonly classified as structural applications. Their more important compositions include brass, nickel silver, and copper-iron powder premixes.

If P/M techniques are used for these components, it is usually because of economic advantages, such as cost savings in labor and materials, over conventional metal-forming techniques such as casting, forging, stamping and machining.

The basic processing steps for the manufacture of these parts are pressing and sintering, including all the variations that are described elsewhere in this book. Naturally, sintering atmospheres, sintering temperatures and other conditions may differ depending on the alloy system being treated.

Brass parts have come into use during the last 20 to 30 years, mainly as a result of the availability of improved brass powders. Applications of brass and nickel silver parts include latch bolts and cylinders for locks; shutter mechanism components for cameras; gears, cams, and actuator bars for timing assemblies and small generator drive assemblies; and decorative trim and medallions. Often, free-machining properties and corrosion resistance are critical to the selection of brass parts.

Commercial prealloyed brass powders are available in leaded and nonleaded compositions. The broad range of useful brass alloys includes 90Cu-10Zn through 65Cu-35Zn; however, leaded versions of 80Cu-20Zn and 70Cu-30Zn are most commonly used for the manufacture of sintered structural parts that may require secondary machining operations.

The only nickel silver powder common to the industry has a nominal base composition of 65Cu-18Ni-17Zn, which is modified only by addition of lead for the purpose of improving machinability.

These classes of alloys are usually blended with lubricants in amounts from 0.5 to 1.0 wt %. Lithium stearate is the preferred base lubricant because of its cleansing, scavaging activity; however, bilubricant systems are common, such as lithium stearate–zinc stearate on a 50/50 basis to minimize the surface staining attributed to excessive lithium stearate.

Both brass and nickel silver compacts sinter to bright matte surface finishes that may be burnished to relatively high lustre for decorative or aesthetic purposes. Most standard prealloyed and lubricated compositions will compact to green densities in excess of 85% of wrought counterpart density at 30 tsi, and, due to normal shrinkage during sintering, final densities in excess of 90% are obtainable. Because of the excellent ductility of these single-phase (alpha) sintered alloys, further movement and densification through cold, postsintering re-pressing results in further general improvements in mechanical properties—particularly yield strength, hardness and dimensional precision.

Sintering of brass and nickel silver compacts is normally performed in protective (nonoxidizing) atmospheres at temperatures ranging from 1500 to 1700°F (816 to 927°C) depending on alloy composition. In all cases, to avoid excessive distortion and/or blistering of the compacts, sintering temperature should not exceed the solidus temperature of the alloy.

Compositions of most alloys are governed by existing specifications and thus may not be varied to any effective degree to significantly alter normal sintered characteristics (MPIF Standard #35 and ASTM Standard B-282). Fortunately, alloys respond rather well to variations in sintering time and temperature, and, consequently, properties such as dimensional change and ductility may be controlled without significant sacrifices in mechanical strength. Relationships among typical sintered properties of two primary alloy systems—70Cu-30Zn and 80Cu-20Zn leaded brasses—are shown in Fig. 16 and 17, respectively. The sintered property relationships of these alloy systems are quite similar. The main differences are in the magnitudes of the property values obtained.

Four examples of brass door-lock components are shown in Fig. 18. Each part was compacted and sintered to a minimum density of 7.6 g/cm^3 and then subjected to required secondary operations such as burnishing, hole tapping and threading. One cylinder is shown fully assembled with key in place. The latch bolt is of a high-security type in which a hardened steel pin is inserted as a mechanical deterrent to forced entry. Figure 19 shows an assortment of nickel silver parts. In many of these applications, corrosion resistance, wear resistance and aesthetic appearance play important roles..

P/M bronze parts for structural applications often are selected because of the corrosion and wear resistance of

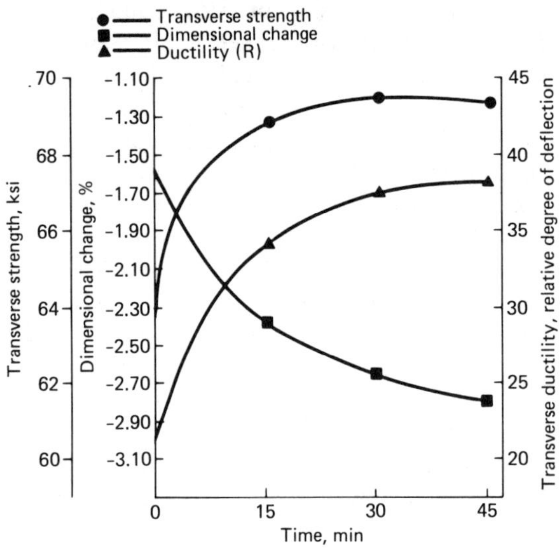

Fig. 16. Mechanical properties of 70/30 leaded brass powder (B-155) vs time at 1600°F. Green density, 7.30 g/cm³.

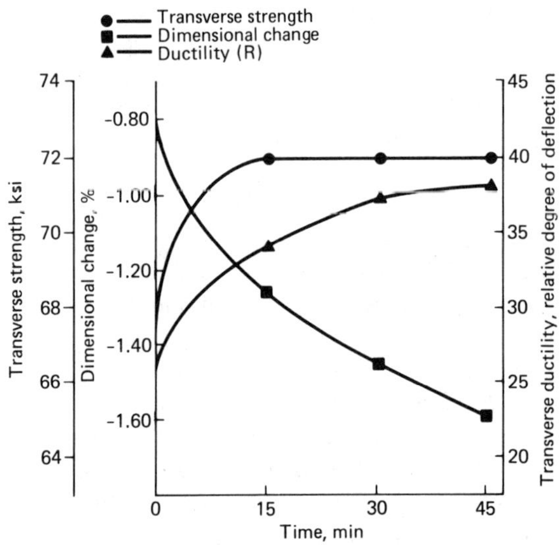

Fig. 17. Mechanical properties of 80/20 leaded brass powder (B-129) vs time at 1600°F. Green density, 7.60 g/cm³.

bronze. The examples shown in Fig. 20 include cams, gears and brake shoes.

Most P/M tin bronze parts are produced from elemental premixes of tin and copper. Typical strength/density curves for 90Cu-10Sn bronzes with and without graphite are shown in Fig. 21. Sintering of these materials requires certain precautions in order to develop optimum microstructures. Prealloyed bronze compositions are available for structural parts requiring densities in excess of 7.0 g/cm^3.

Copper-nickel powders containing 10 and 25% nickel have been used for coinage and corrosion-resisting applica-

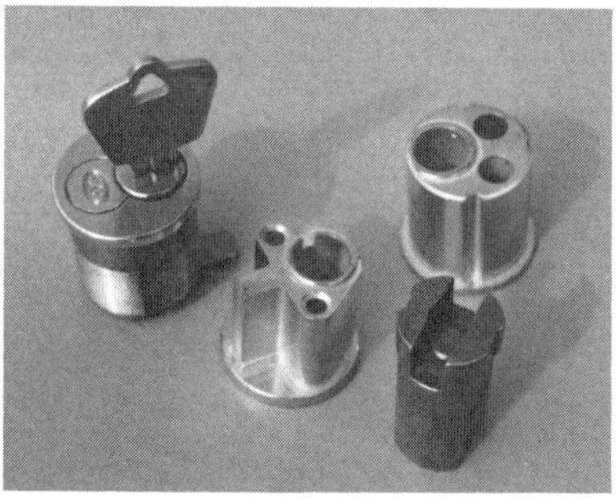

Fig. 18. P/M brass door-lock components

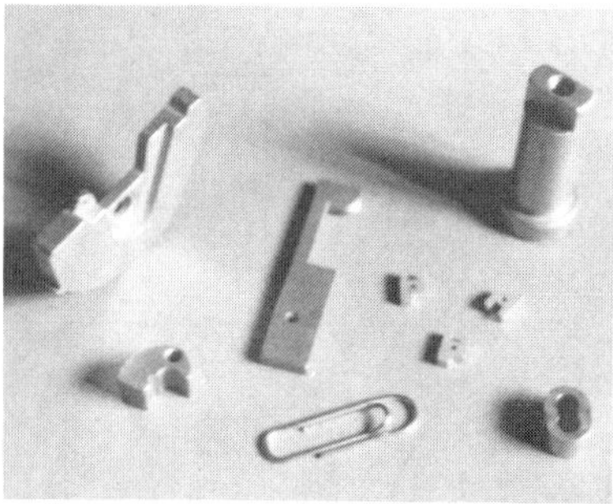

Fig. 19. Assorted P/M nickel silver parts

Fig. 20. Assorted P/M bronze parts (Courtesy of Norddeutsche Affinerie)

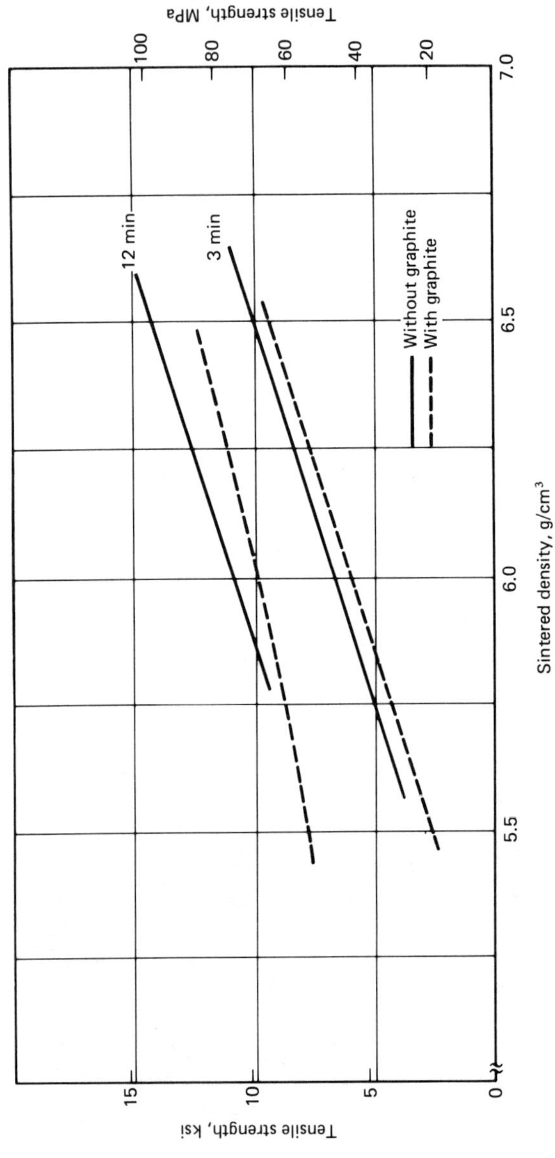

Fig. 21. Effect of density on strength of 90/10 bronze compacts with and without graphite

tions. No standards exist for these materials, but processing instructions and property data can be found in the literature.

Copper powder is added to iron and steel powders to increase mechanical properties. Two approaches are used. In one, copper powder (up to about 10%) is added to the steel powder prior to pressing and sintering; sintered densities range from about 80 to about 90% of theoretical. Typical strength/density curves for such a material (5% Cu, 0.8% C) are shown in Fig. 22. In the other approach, copper is added through an infiltration process which essentially produces a fully dense part (see "Copper-Infiltrated Parts," below).

Electrical Parts

The use of pure copper P/M parts, mainly in electrical and electronic applications, is still not very widespread even though the processing procedures and property data required for obtaining high electrical conductivity have been made available within the past 10 years. It is essential either to use very pure copper powders or to bring about the precipitation of soluble impurities during sintering.

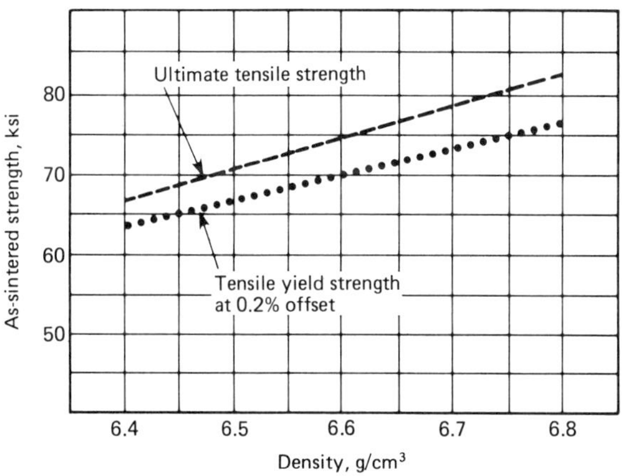

Fig. 22. Strength/density curves for P/M copper steel (5Cu-0.8C) MPIF standard

If these precautions are taken it is possible to obtain the values of strength and conductivity shown in Fig. 23. For a given porosity, the electrical conductivity of a porous copper part depends on pore characteristics: small, spherical pores are preferable to large, elongated pores. Full-density properties are reached or approached by double pressing, coining or forging. Typical applications include commutator rings, contacts, shading coils, nose cones and electrical twist-type plugs.

A new copper product made by a P/M technique is quite different from the above-described P/M copper and is gaining increasing acceptance. It is an aluminum oxide dispersion-strengthened copper, available under the trade name "GlidCop". This material is made by internal oxidation of a copper-aluminum powder. The fine dispersion of aluminum oxide generated during internal oxidation retards recrystallization. Thus, the material can withstand high-temperature exposure without appreciable softening. The fully consolidated material is available in a wide range of sizes of rod and other cross sections, in strips of various thicknesses, and in wire. Property data are shown in Fig. 24 to 27. The material has a combination of good electrical conductivity (from 78 to 92% IACS depending on dispersoid content) and excellent elevated-temperature stability. This makes it useful in elevated-temperature applications and where retention of these properties after exposure to elevated temperature is important. It also may be drawn into fine wire or rolled into strip without intermediate annealing. A major application is electrode caps for non-consumable resistance welding, for which this material has outperformed RWMA Class 2 type welding electrodes by a factor of six (Table 11 and Fig. 28).

In another application, GlidCop is used for the leadwires to the tungsten filaments in light bulbs (Fig. 29). This is of some historical interest due to the fact that the tungsten filament was probably the first (although unintentional) commercial application of a dispersion-strengthened material, the performance of which is now further improved by the addition of dispersion-strengthened copper.

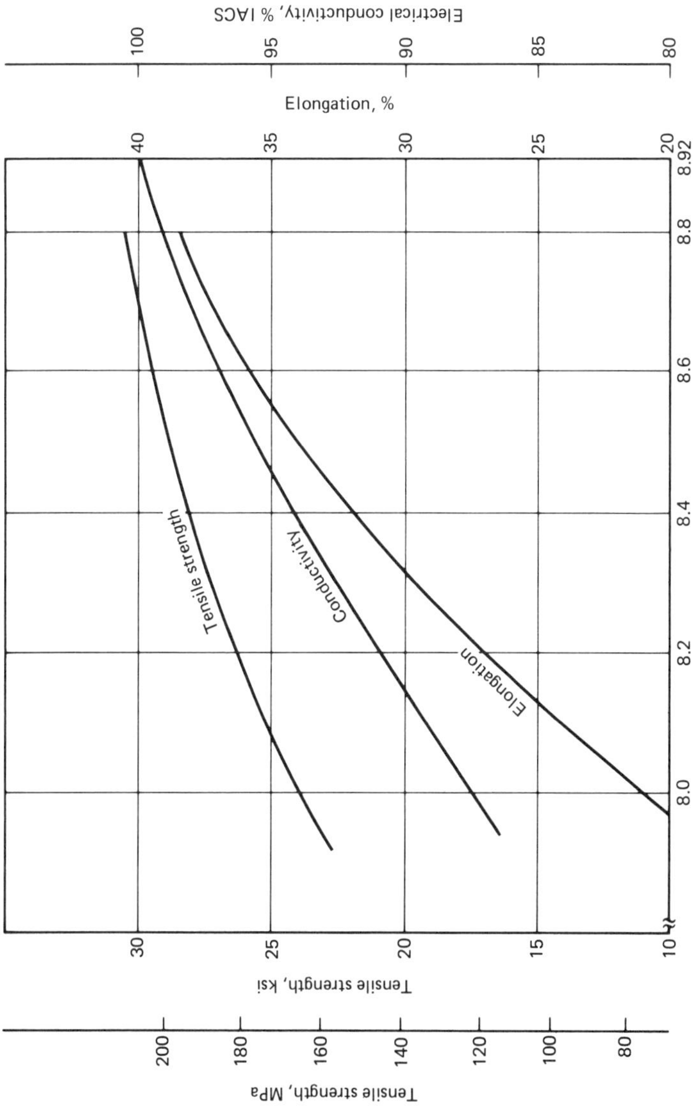

Fig. 23. Effect of density on electrical conductivity of P/M copper (Ref. 4)

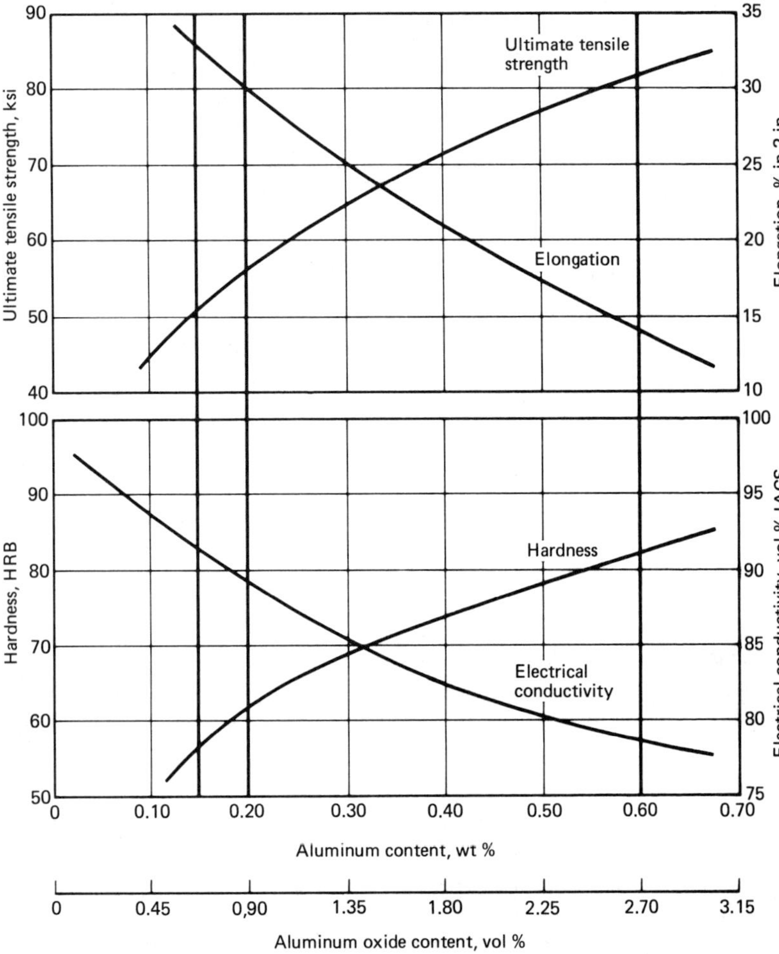

Fig. 24. Properties of GlidCop

Copper-Infiltrated Parts

In copper infiltration the pores of a sintered part are filled with liquid copper. Attempts to obtain full density in a porous P/M steel part through copper infiltration date back to 1930. Because of poor surface appearance and a lack of dimensional control, these attempts did not lead

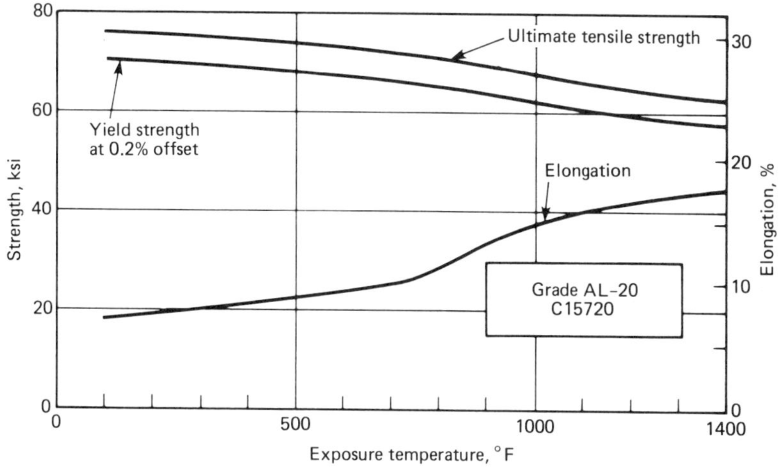

Fig. 25. Effects of 1 h at elevated temperature on room-temperature mechanical properties of 0.032-in. AL-20 GlidCop strip

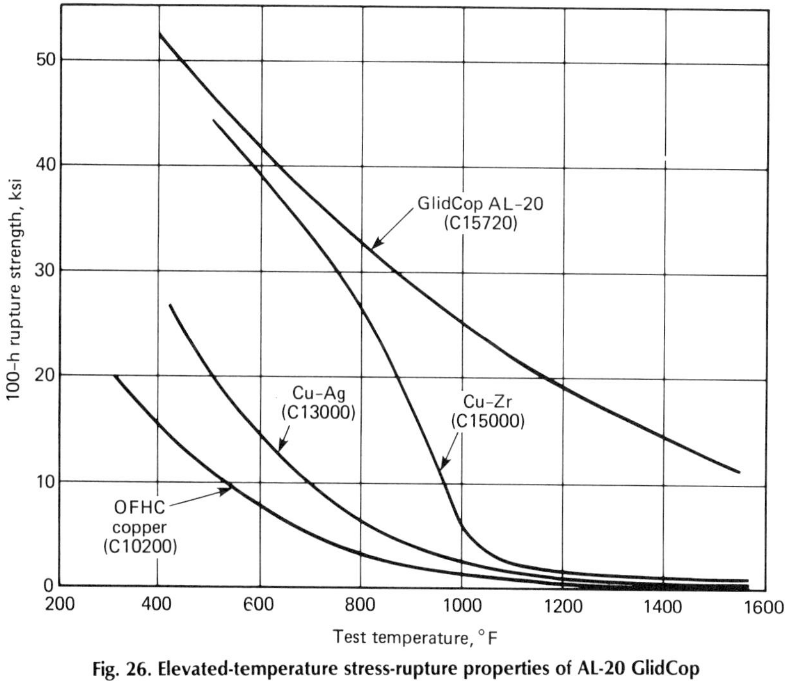

Fig. 26. Elevated-temperature stress-rupture properties of AL-20 GlidCop

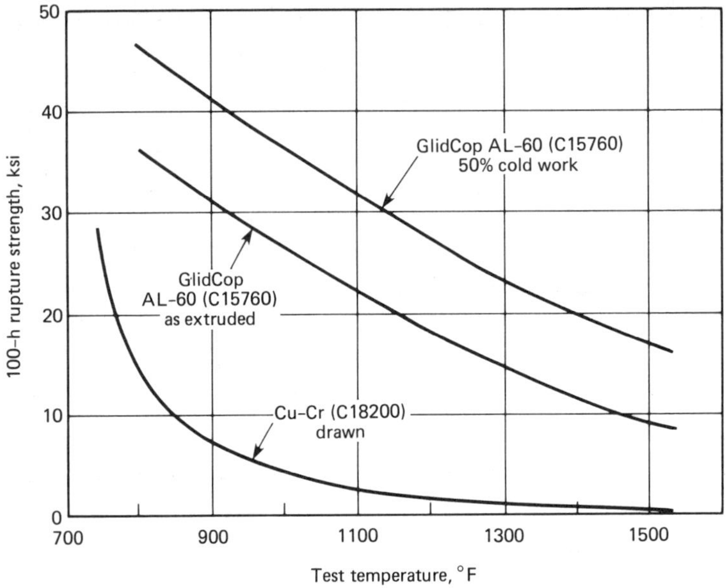

Fig. 27. Elevated-temperature stress-rupture properties of AL-60 GlidCop

Fig. 28. Assorted GlidCop welding electrode caps

to commercial success. In 1946 these problems were solved through addition of graphite to the P/M steel part and addition of iron to the copper infiltrant, which prevented erosion of the P/M steel part.

Table 11. Welding conditions for various applications of GlidCop electrodes

Application	Force, lb	Current, A	Cycle time	Welds per hour	Thickness(a) and type(b) of material welded	Electrode life(c)
Rocker panel	600	15,000	15	3024	0.031 G/0.031 G	10 to 1
Rocker panel	1080	15,500	15	2600	0.054 G/0.068 G	5 to 1
Light truck box	800	16,000	18	160	0.043 G/0.054 CR	6.5 to 1
Pillar assembly	700	14,000	13	228	0.043 G/0.054 G	6 to 1
Cowl side to rocker	980	15,500	13	1750	0.050 CR/0.053 G	5 to 1
Rocker panel to body	980	15,000	13	1720	0.054 CR/0.068 G	6 to 1
Floor panel extension	550	13,500	12	645	0.031 CR/0.031 G	4.6 to 1
Windshield header	850	14,000	13	645	0.036 CR/0.068 CR	6 to 1
Dash extensions	550	13,500	12	1144	0.031 CR/0.031 CR	5.5 to 1
Station wagon assembly	660	12,000	9	3120	0.054 CR/0.031 CR	4 to 1
Van assembly	775	13,500	18	180	0.043 CR/0.056 CR/0.036 G	6 to 1
Wheelhouse assembly	770	12,500	11	3300	0.031 Z/0.033 Z	4 to 1
Dash panel	550	13,500	12	1806	0.031 Z/0.068 G	6 to 1
Wheelhouse assembly	630	12,265	10	546	0.031 Z/0.031 CR	8 to 1

(a) In inches. (b) G = galvanized steel; CR = cold rolled steel; Z = zincrometal. (c) Life of GlidCop electrodes compared to life of standard class 2 material electrodes.

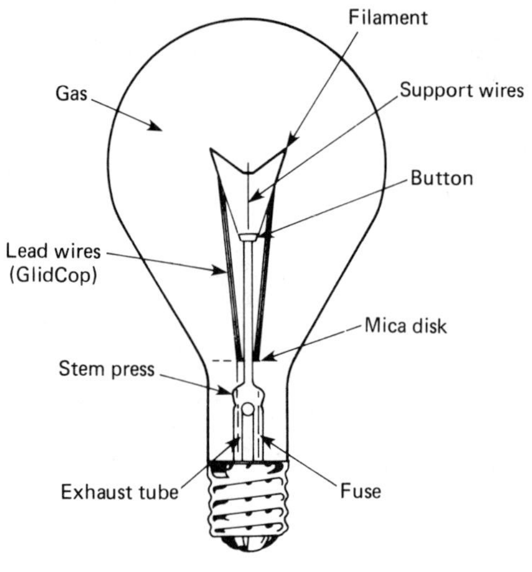

Fig. 29. GlidCop lead wires in a light bulb

The major developments in this field since World War II may be summarized as follows:

- New copper infiltrants offering improved performance have been developed.
- Larger parts are being infiltrated.
- The average matrix density of the steel part is higher because of the availability of inexpensive high-compressibility iron powder.
- Water-atomized hardenable low-alloy steel powders may be infiltrated.

A pictorial version of infiltration is shown in Fig. 30, which illustrates the widely used one-step infiltration process (also known as "sintration") in which sintering and infiltration of the base part are combined into one processing step. The infiltrant slug may rest either on top of the base part (top-infiltration) or beneath it (bottom-infiltration). Typical infiltrating conditions are 30 minutes at 2050°F in endo gas with dew points anywhere from 20 to 50°F. Dissociated ammonia or nitrogen-base atmos-

Right: green P/M steel part with copper infiltrant slug. Center: after sintration. Left: after removal of residue.

Fig. 30. Three stages in copper infiltration of P/M steel part

pheres are also used. There are so-called residue-free and residue-type infiltrants: the latter leave a small nonadhering residue which can be disposed of easily.

Figures 31 and 32 show assortments of copper-infiltrated parts. These parts are widely used in load-bearing applications in many types of machines and engines. P/M gears are particularly well suited to copper infiltration, because infiltration adds strength to the teeth of the gear, which often suffer from lower density.

Fig. 31. Assorted copper-infiltrated P/M steel parts

Fig. 32. Assorted copper-infiltrated P/M steel parts

The single most important reason for infiltrating a part with copper is to enhance its mechanical properties. Sometimes, however, other properties, listed in Table 12, are important either alone or in combination with higher strength.

Figure 33 shows the effect of infiltration on the transverse rupture strength (TRS) of a steel part. For Ancorsteel 1000 infiltrated to 7.4 g/cm^3 with about 20% copper, the TRS after infiltration is about twice that without infiltration, with a corresponding increase in density of about 15%.

Figure 34 shows strength as a function of the amount of void volume infiltrated. Much of the commercial infiltration is done in the upper range (80 to 100%).

Table 12. Uses of copper infiltration

Enhancement of mechanical properties
Higher and more uniform density
Sealing of porosity for secondary operations (pickling, plating, brazing, painting)
Selective properties through localized infiltration
Multiple-component assemblies
Improvement of machinability
Improvement of conductivity
Noise level reduction in moving components

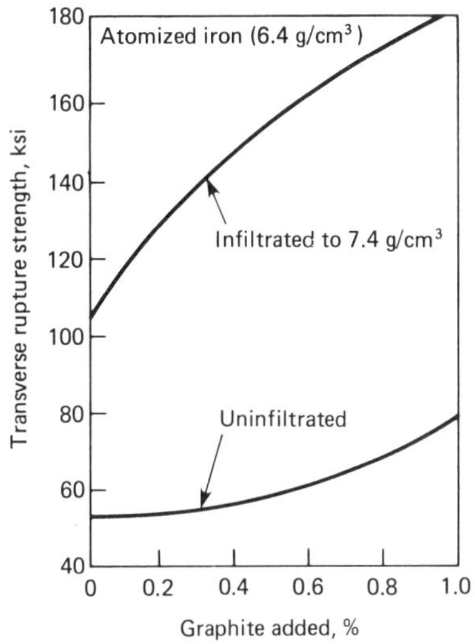

Fig. 33. Transverse rupture strength of copper-infiltrated steel part as a function of carbon content

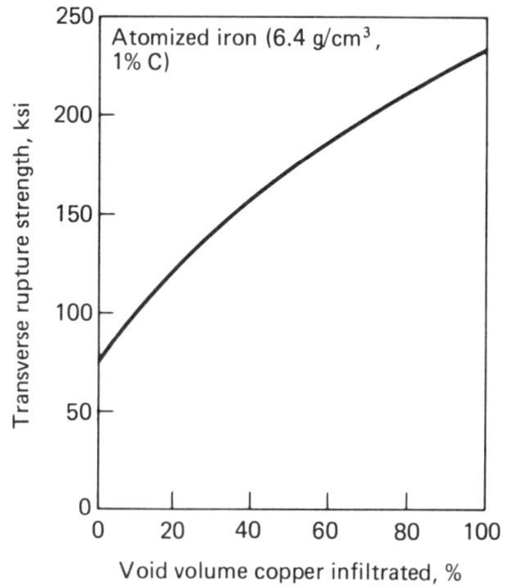

Fig. 34. Transverse rupture strength of copper-infiltrated steel part as a function of void volume infiltrated

Parts producers usually provide complete sets of data on properties of infiltrated parts, including dimensional tolerances. Standard specifications are available from MPIF and other organizations. Table 13 shows one of these standards. Infiltrant powder producers often provide the kind of data that are characteristic of their materials and that facilitate their use for the parts producer.

Figure 35 shows the typical pearlitic microstructure of a copper-infiltrated iron-carbon part. Austenitizing, quenching and tempering produces tempered martensite. Rounding of the iron grains, which is due to dissolution and reprecipitation of iron, improves ductility and impact strength. Furthermore, due to mutual solubility during infiltration but little room-temperature solubility, precipitation strengthening occurs in both matrices.

Successful infiltration requires that the surface free energy of the system be smaller after infiltration than

Table 13. MPIF standard for copper-infiltrated P/M steel

P/M MATERIALS STANDARD **APPROVED: 1972** REVISED:

Material	Designation
P/M Infiltrated Steel	FX–1005–T

1	Chemical Composition (per cent)	Element	Fe	Cu	C						
		From	80.5	8.0	0.3						
		To	91.7	14.9	0.6(1)						
			Other Elements: Total by difference equals 4.0 max.								

Line / Col.	1		2	3	4
2	Condition		As-Sintered	Heat Treated	
3	Density	g/cc	7.2 – 7.6	7.2 – 7.6	
4	Allowable Density Variation	g/cc	0.4	0.4	
5	Interconnected Porosity	%			
6	Properties, Mechanical		Note: Typical properties based on reference tests (see Properties page 4, Standard 35)		
7	Ultimate Tensile Strength (See line 23)	psi	83,000	120,000	
8	Tensile Yield Strength 0.2% Offset (See line 23)	psi	64,000	107,000	
9	Compressive Yield Strength 0.1% Offset	psi			
10	Elongation	%	4.0	1.0	
11	Apparent Hardness		Rb 75	Rc 35	
12	"K" Strength Constant				
13	Impact Strength (unnotched charpy)	ft.-lbs.	14.0	7.0	
14	Fatigue Strength	psi			
15	Special Properties				
16	Typical Applications		Mechanical Components		

Table 13 continued

Line / Col.	1		2	3	4
17	Properties (Continued)		As-Sintered	Heat Treated	
18	Young's Modulus	psi	20.0×10^6	20.0×10^6	
19	Poisson's Ratio				
20					
21					
22					
23					
24	Comments: Usage limited to special part configuration. Consult P/M part manufacturer before specifying. (1) Combined carbon, on the basis of iron only, may be a metallographic estimate.				

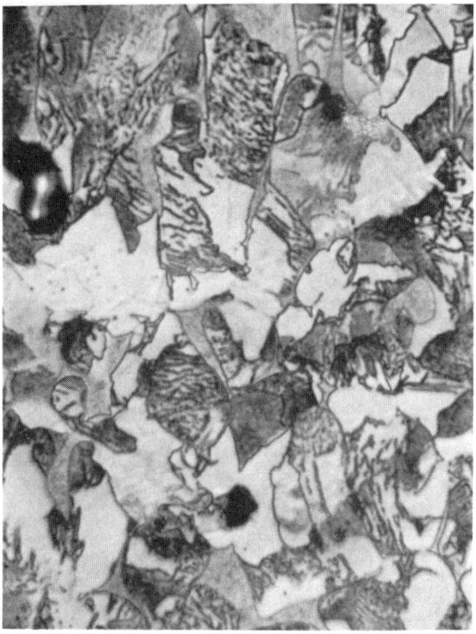

Fig. 35. Pearlitic microstructure of copper-infiltrated steel

before it. Infiltration theories were formulated by Semlak and Rhines in 1958, and by Shaler in 1965.

Other Applications

Other applications that are not specifically covered above but that complement the usage of copper and copper alloys are briefly reviewed as follows: Flaked (ball milled) and other forms of copper may be used in combination with graphited materials to form "carbon brushes" used extensively in electrical motor units. Other uses include industrial coatings, metallic greases, thermal control heat-sensing devices, and decorative paints. Copper-base alloys such as brasses are also manufactured in flake form and are typically used in decorative finishing, such as in paints, plastics, and other various forms of printed articles. Miscellaneous uses include fillers in brazing compounds; in

PTFE compounds (bronze); and in epoxies and polyester-type resins (brass and bronze).

The preceding review illustrates the versatility and very broad application of copper and copper-base alloy powders. This group of nonferrous materials represents a vital portion of the powder metallurgy industry as related to the fabrication of metallic P/M structural components and satisfies other major industrial and commercial needs.

Selected References

1. P. W. Taubenblat (Ed.), *Copper Base Powder Metallurgy*, Vol 7, New Perspectives in Powder Metallurgy, Metal Powder Industries Federation, Princeton, NJ, 1980
2. *Copper and Copper-Base Powder Alloys* (a literature survey of recent technical advances), Metal Powder Industries Federation, Princeton, NJ, 1976
3. F. V. Lenel, *Powder Metallurgy – Principles and Applications*, Metal Powder Industries Federation, Princeton, NJ, 1980
4. J. L. Everhart, *Copper and Copper Alloy Powder Metallurgy: Properties and Applications*, Metal Powder Industries Federation and Copper Development Association, n.d.
5. Werner Schatt (Ed.), *Pulvermetallurgie Sinter- und Verbundwerkstoffe*, VEB Deutscher Verlag für Grundstoffindustrie, Leipzig, 1977
6. J. S. Hirschhorn, *Introduction to Powder Metallurgy*, American Powder Metallurgy Institute, Princeton NJ, 1969

Chapter 7

The Effects of Phosphorus on the Properties of Iron Alloys for Magnetic Applications

Kenneth H. Moyer
Hoeganaes Corporation

Powder metallurgy has made vast inroads into technology for development of better soft magnetic materials for the electronics era. Five years ago, even though iron powders had been used to make parts such as pole pieces since the early 1950's, not much was understood about how porosity affected the properties of these materials. P/M magnets were relegated to those applications requiring only limited properties.[1,2]

Today iron P/M magnets can be produced with properties that equal those of some of the best iron magnets produced commercially by any technology.[3] Furthermore, these properties can be predicted more accurately than can those of most iron magnets produced commercially. Porosity, long touted as a deterrent to good properties, can be controlled, and excellent structure-sensitive properties can result.

Still, many problems beset powder producers. Only a few can meet the stringent properties demanded for the best magnets. Most still resist the belief that carbon, nitrogen and oxygen cause structure-sensitive properties to be impaired. Others still insist on limiting their designs to silicon iron or iron.

Bozorth, in the early 1950's, reported properties of phosphorus irons and pointed out several advantages of using these materials in magnetic devices. Unfortunately, phosphorus renders iron or steel hot short, and so the advantages of alloying with phosphorus have not been exploited.[4] This need not be the case when P/M techniques are employed, however, because production of P/M products does not require hot working, and therefore the advantages stemming from addition of phosphorus to powders can be realized. Even in the short period of time that has elapsed since interest in the use of powders in the electronics industry was rejuvenated, much has been published on the properties of phosphorus iron,[5-7] depicting the benefits described so long ago by Bozorth. The Japanese have also extended the use of silicon iron by showing that up to 6% silicon can be added to P/M compacts to further improve their resistivity.[8] Although not much is being published, the P/M community is taking its first glimpse of powder products for AC applications.[9] Even at this early stage, people are starting to think about how powders can be used for this vast untapped market.

The purpose of this chapter is to provide information which will show the advantages that can be realized by producing magnets from a commercially available 0.45% phosphorus iron. It provides an explanation of how porosity can be controlled in order to develop optimum magnetic properties in this alloy. It also introduces predictor equations which permit designers to fabricate quality parts from P/M compacts. Finally, it points out some of the advantages of alloying phosphorus with iron to improve magnetic properties and demonstrates how properties can be compromised if pure powders and best sintering practice are not employed.

The Effects of Phosphorus on DC Properties

Whenever designers start to consider alloys for DC magnetic applications, one of the first considerations is the effect of alloying on saturation. When 0.45% phosphorus is added, and no porosity is present, the saturation is reduced by approximately 500 gausses. This is a small price to pay in order to increase resistivity by almost 100%. The effect of adding phosphorus to iron is shown in Fig. 1. It can be readily seen that saturation is only marginally decreased.

The effect of porosity on saturation is illustrated in Fig. 2. As density decreases, saturation is lowered, and the applied field required for achieving saturation becomes greater. The engineer using these porous materials can simply modify the saturation according to the chosen density and accommodate the greater field required with a

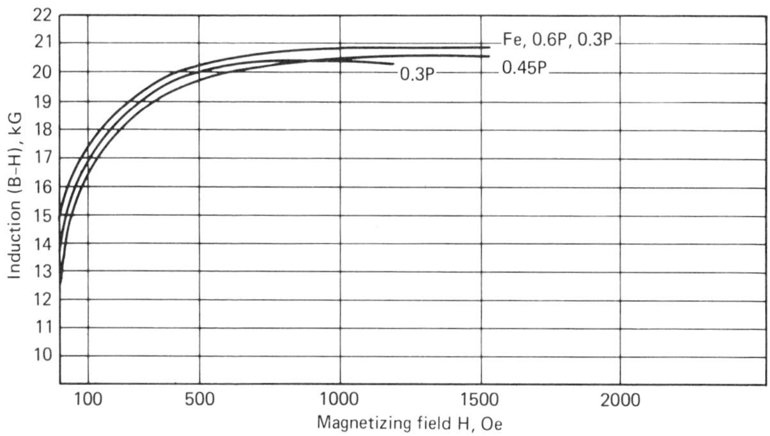

Fig. 1. Magnetization curves for hot re-pressed iron-phosphorus alloys

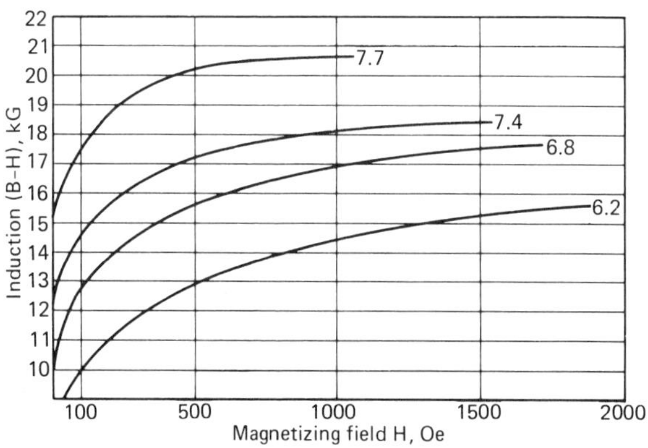

Fig. 2. Magnetization curves for atomized iron at various densities

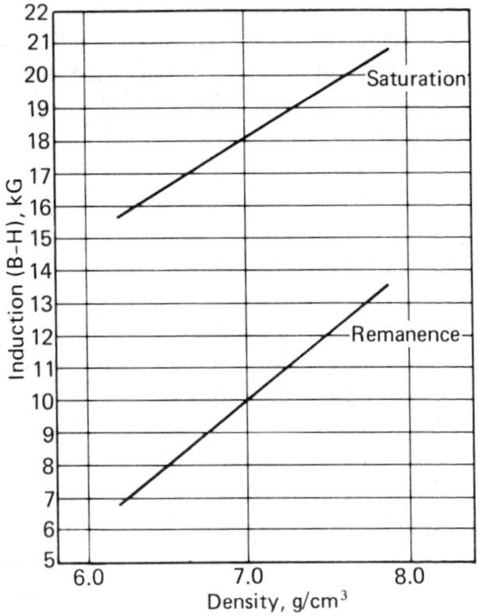

Fig. 3. Magnetization of iron as a function of density

change in ampere turns. Since saturation increases linearly with density (Fig. 3), it is easy to design for saturation simply through pressing and sintering to the proper density.

Full-Density Parts

It has been shown that when no porosity is present in P/M parts, their magnetic properties are like those of any other iron parts.(3) Bozorth has further shown that, if phosphorus is added, permeability as well as resistivity are increased and coercive force is lowered.(4) If greater sensitivity is required, or if the material may be applied under other than DC conditions, these advantages may be realized. Before the effects of porosity can be understood, it is important to understand how other factors affect the properties of fully dense parts made of phosphorus-containing iron.

Production of modern atomized powders employs the best steelmaking practice known today. The result is that extremely pure iron can be produced through this technology. Bozorth again points out that, if exceptional magnetic properties are to be achieved, the material must be pure.[4] Purity results in low residual strain of the characteristic crystal lattice structure. Since powder particles oxidize more than do conventional materials, their oxygen content is higher, and oxygen compared with carbon, nitrogen and sulfur causes the least deterioration of structure-sensitive properties. Also, most of the surface oxides are reduced during sintering, so these properties are moderately changed at most. The effect of sintering temperature on oxygen content is shown in Fig. 4. If high-temperature sintering is employed, oxygen contents of powders can approach those of wrought materials. Carbon, nitrogen and sulfur may also be maintained at extremely low levels provided that good melting practice and careful sintering of compacts are employed.

The second factor affecting structure-sensitive magnetic properties is the grain size of the material. For best structure-sensitive properties, permeability and coercive force, grain growth must occur, and the larger the grains the better. Grain growth is also affected by sintering practice (Fig. 5). It can be seen that if high-temperature

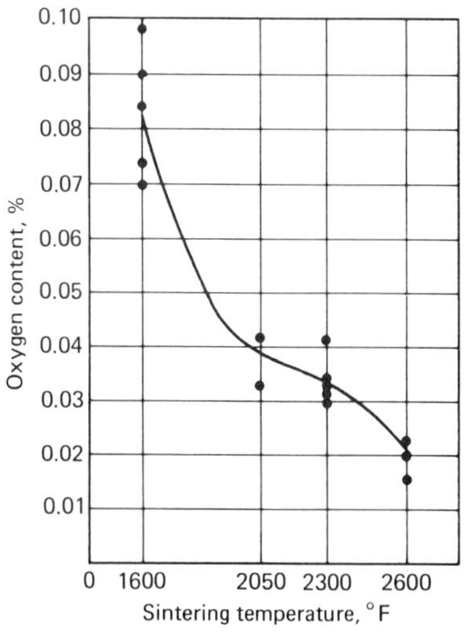

Fig. 4. Effect of sintering temperature on oxygen content of iron

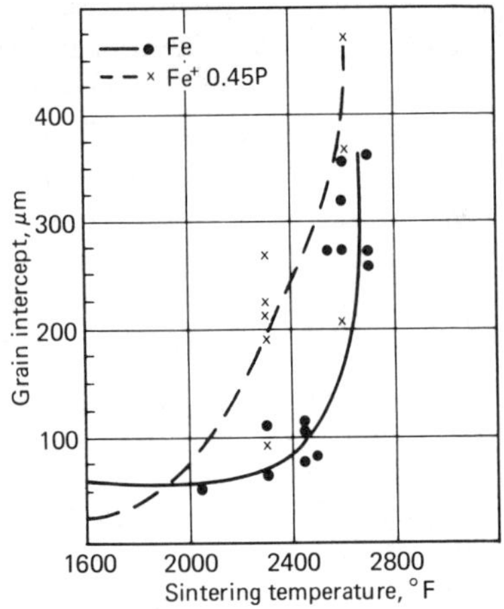

Fig. 5. Variation of grain intercept with sintering temperature

sintering is employed grain growth occurs, just as in any other material. In the case of iron, significant grain growth does not occur until the material is sintered above the delta transition temperature, when large grains are achieved. Phosphorus (0.45%) provides an advantage in that a liquid phase enhances grain growth so that at temperatures as low as 2300°F (1260°C) grain growth occurs.

Residual induction is primarily a linear function of density, but at any density level it can be lowered as the purity of the material is increased. Both permeability and coercive force improve as purity improves and as grain growth occurs.

The effects of these two variables on remanent magnetization are shown in Fig. 6. If conventional sintering temperatures are employed, the oxygen contents of the compacts are at the highest levels and virtually no grain growth occurs. The result is high levels of residual induction. When high-temperature sintering is employed, oxygen is reduced and grain growth occurs. The result is a lowering of the residual induction to levels more commensurate with those of pure wrought materials. In the case of iron, remanent magnetization was found to be invariant.[3] A phosphorus addition of 0.45% permits this property to vary. Since it can be predicted it may be used for design by simply regulating the oxygen content (Fig. 7). This is accomplished through rigorous control of sintering conditions. It is important in the design of solenoids, because the forces for proper release of these devices are rigorously controlled by the remanent magnetization.

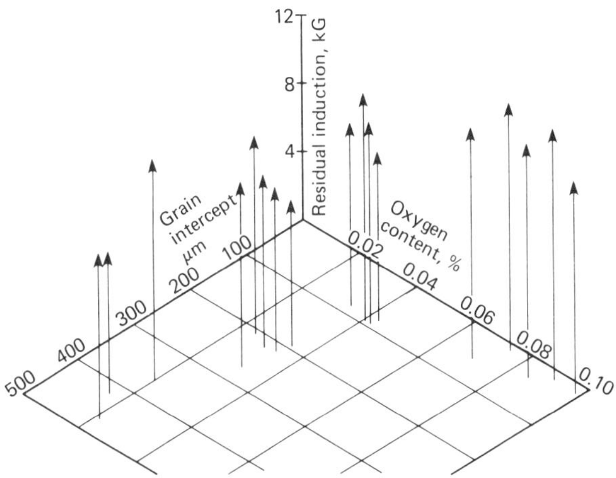

Fig. 6. Effects of oxygen content and grain intercept on residual induction of hot re-pressed Fe-0.45P compacts

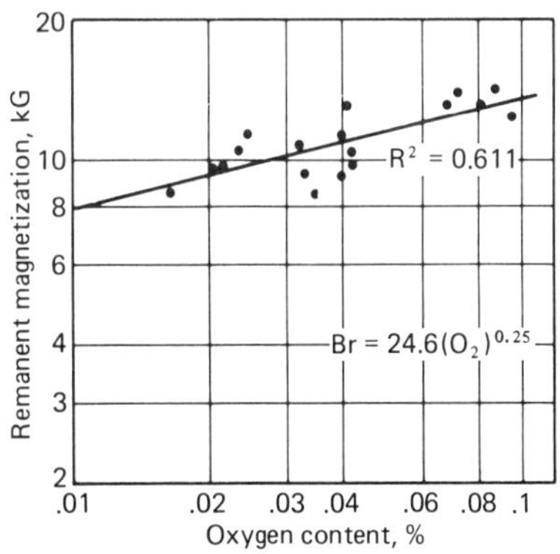

Fig. 7. Predictor equation for remanent magnetization as a function of oxygen content

The same is true of the other structure-sensitive properties. Figure 8 shows the effects of oxygen content and grain-size intercept on maximum permeability. If oxygen content is high and no grain growth has occurred, maximum permeability is at its lowest levels. Once the oxygen has been reduced and grain growth has occurred, maximum permeability reaches optimum levels. Bozorth's prediction of the beneficial effect of phosphorus on magnetic properties is also demonstrated in this figure.[4] Here, data are plotted for iron and for iron containing 0.45% phosphorus for the different sintering conditions. When phosphorus is added, maximum permeability is significantly increased. Others have found even more beneficial effects by adding up to 0.8% phosphorus.[10] It was previously shown that the residual induction of iron can be readily predicted as a function of the grain-size intercept.[3] This equation, with its two sigma limits, is shown in Fig. 9. The data points for the 0.45% phosphorus iron are also shown. It can be seen that these points lie on the upper two sigma line for the iron predictor equation.

The same relationship holds true for coercive force. Figure 10 shows that, as oxygen is reduced, grain growth

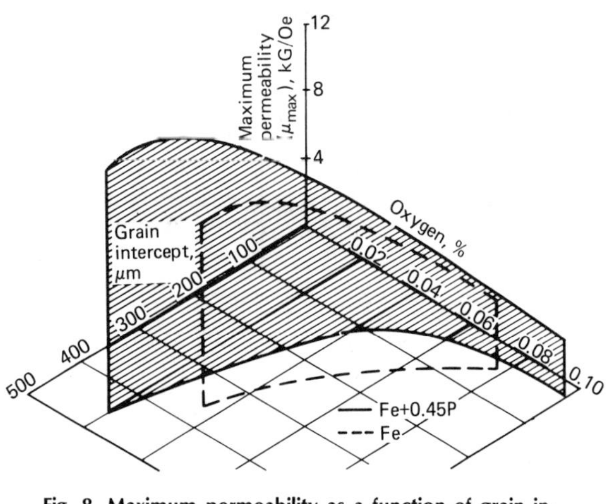

Fig. 8. Maximum permeability as a function of grain intercept and oxygen content

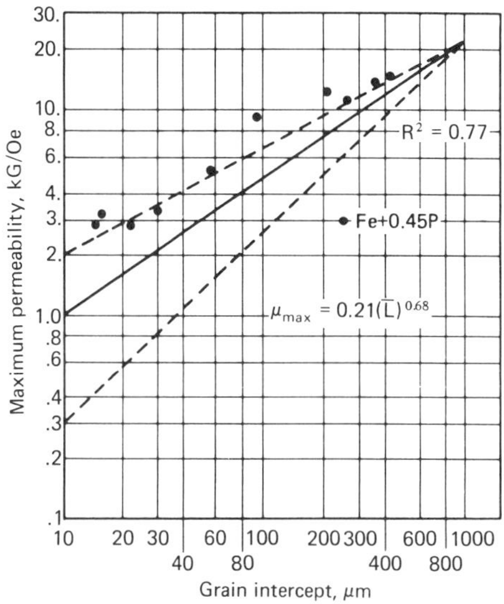

Fig. 9. Comparison of maximum permeability of hot repressed Fe-0.45P with predictor equation for iron as a function of grain intercept

occurs and low coercive force is achieved. This is true for both iron and the 0.45% phosphorus iron alloy. A predictor equation for iron was also developed and is shown with its two sigma limits in Fig. 11. The data points for the 0.45% phosphorus iron alloy are seen to lie almost directly on the equation for iron, with almost no deviation.

Porous Compacts

Although P/M parts can be made fully dense by hot working processes such as forging or extruding, most P/M parts are not produced beyond density levels of 7.4 g/cm^3. It is therefore important to understand how magnetic properties are affected if pores are present in the part. For this purpose, cores were compacted from powders possessing narrow size fractions to various density levels and sintered at temperatures ranging from 1600°F (870°C) to 2600°F (1430°C). By keeping the size fractions of the

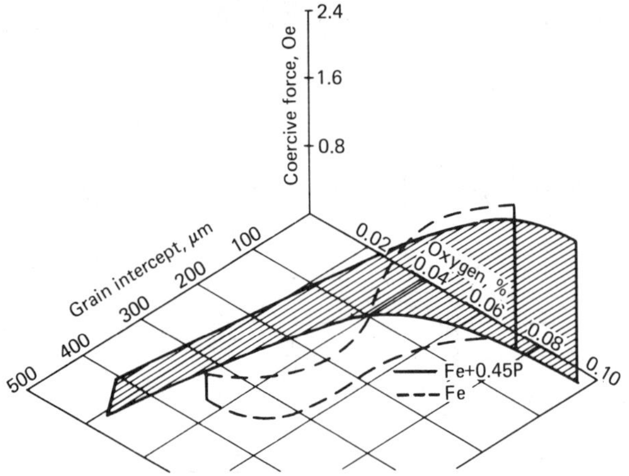

Fig. 10. Coercive force as a function of grain intercept and oxygen content

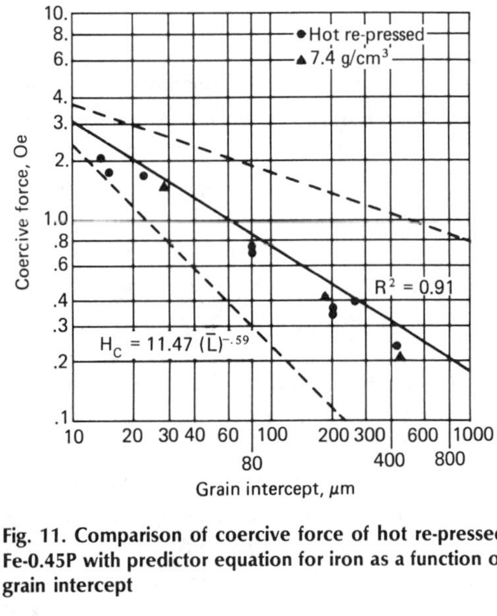

Fig. 11. Comparison of coercive force of hot re-pressed Fe-0.45P with predictor equation for iron as a function of grain intercept

particles within a narrow range, coarse or fine pores could be introduced within the part. Similarly, by sintering at various temperatures the areas of the pores could be made large or small and the pores could be rounded. Also, compacting to various densities caused the pores to be close together or spaced farther apart. The effect of this was determined by measuring the distance between the pores and the area of the pores using quantitative image analysis. Other independent variables included in the analysis were oxygen content, nitrogen content, density, and mean-grain-size intercept. Dependent variables were induction, measured at 15 oersteds (1190 A/m), residual induction, maximum permeability, coercive force, and resistivity. A regression analysis was made for each density level. The prime independent variable was then selected and a further regression analysis was made over all densities investigated. In some cases two regression analyses were made.

In the case of induction, it has already been determined for iron that this was strictly controlled by density.[3] The predictor equation is shown in Fig. 12 with its two sigma limits. Regression analysis of samples ranging in sintered

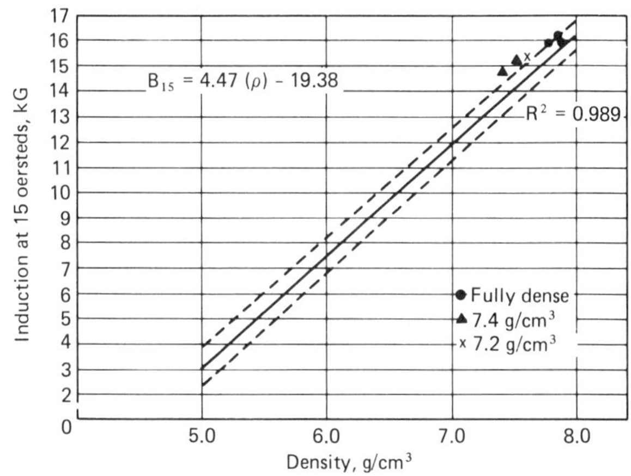

Fig. 12. Comparison of effect of density on induction of Fe-0.45P compacts with predictor equation for iron compacts

density from 7.2 g/cm^3 to full density proved this to be true for the 0.45% phosphorus iron alloys as well. These data points are superimposed on the graph. It can be seen that they also follow the predictor equation that was previously determined for iron.

If porosity is present, density controls the residual induction of phosphorus iron just as it does for iron.[(3)] At any density level, however, core purity and pore size are of secondary importance. If pores are large, remanent magnetization can be lowered by as much as 2 k gausses. Pores also coalesce and spheroidize when compacts are sintered at higher temperatures. As the pores become larger, the residual induction is lowered (Fig. 13). In this figure two linear equations for compacts pressed to densities of 7.4 and 7.2 g/cm^3 are shown with residual inductions as a function of the mean pore area. Note that the slopes are the same but that the equation for the compacts pressed to a density of 7.2 g/cm^3 is lower than the one representing compacts pressed to 7.4 g/cm^3. Since there was less material to magnetize, the residual induction was naturally lower. When the pores were large, however, the residual induction was significantly lower. This provides a further advantage for 0.45% phosphorus iron, because not only can the density be used to advantage, for design purposes, but also, by using coarse powders containing large pores, the residual

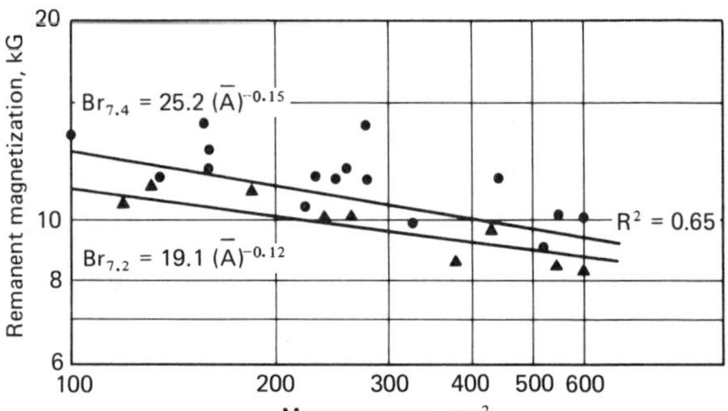

Fig. 13. Residual induction as a function of pore area for Fe-0.45P compacts at densities of 7.4 and 7.2 g/cm^3

induction can be further controlled. Improved solenoids with better control become possible.

The maximum permeability of iron compacts was found to depend on grain size.(3) It was believed that higher oxygen contents within traditionally sintered compacts restricted grain growth. In the case of the 0.45% phosphorus iron compacts, this is also true. The predictor equation for iron is shown in Fig. 14.

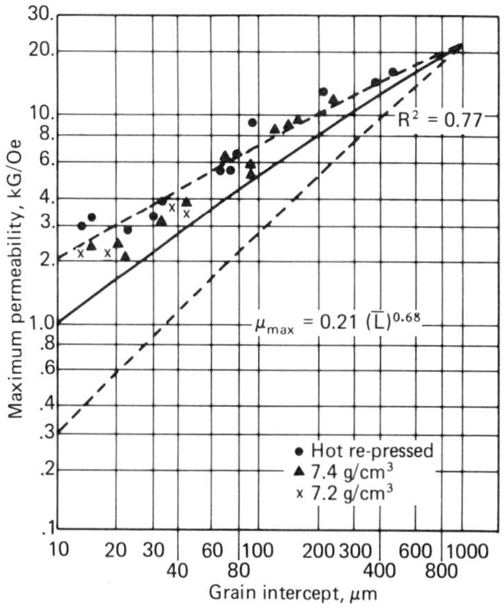

Fig. 14. Comparison of maximum permeability of hot re-pressed Fe-0.45P at densities of 7.4 and 7.2 g/cm^3 with predictor equation for iron as a function of grain intercept

The data points for the compacts pressed to densities of 7.2 and 7.4 g/cm^3 and sintered at the various temperatures are also shown. These fall on the upper two sigma line for the iron predictor equation, so this line can be used to predict this property as a function of grain size.

The same effect was found for coercive force since both are structure-sensitive properties. The predictor equation for iron compacts for this property is shown in Fig. 15.

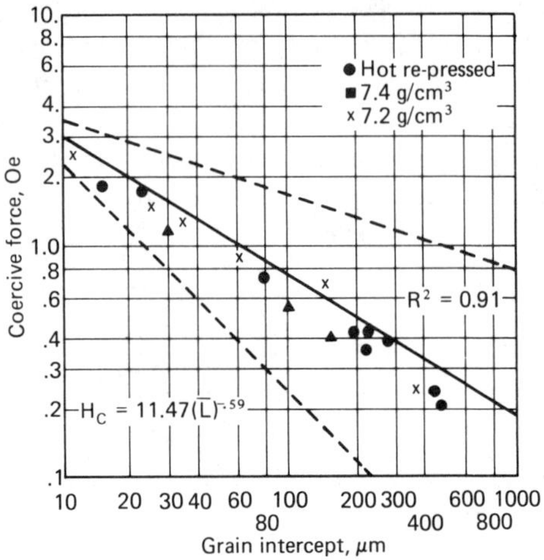

Fig. 15. Comparison of coercive force of hot re-pressed Fe-0.45P at densities of 7.4 and 7.2 g/cm^3 with predictor equation for iron as a function of grain intercept

Again, the data points for the sintered 7.4 and 7.2 g/cm^3 0.45% phosphorus iron compacts are plotted in relation to the predictor equation. The data points fall close to the line for the iron compacts. A further regression analysis of this property was made using oxygen content as the independent variable. The predictor equation is shown in Fig. 16. It can be seen that coercive force may be expressed as a linear function of oxygen content. This is a further advantage of using the 0.45% phosphorus iron alloy. Most parts producers have equipment for determining the oxygen contents of sintered compacts. It is unlikely that they would be disturbed if oxygen content were included as part of a specification. Most, however, are unfamiliar and uncomfortable with grain-size measurement techniques and would therefore be reluctant to include grain size as part of a material specification. Since maximum permeability and coercive force are closely related, oxygen control of compacts may be used to design and control the properties of those magnetic parts that require optimum structure-sensitive properties.

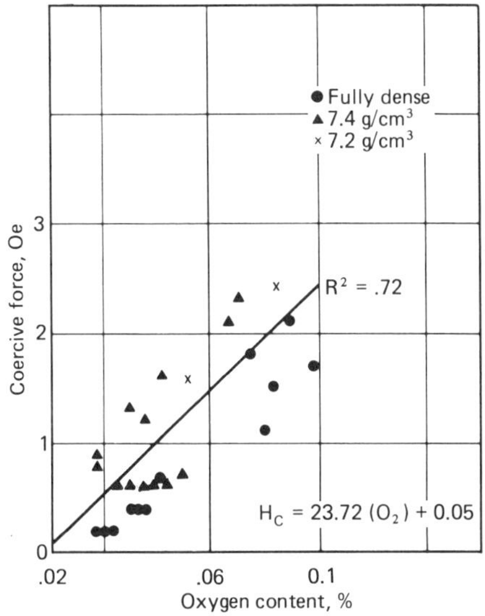

Fig. 16. Influence of oxygen content on coercive force

Resistivity is also enhanced by addition of phosphorus to iron. It is known that phosphorus increases the resistivity of iron.[4,6,7] If 0.45% phosphorus is added, the resistivity is almost doubled. In many instances this increase in resistivity has made it possible to use this alloy for magnetic parts operating intermittently under AC conditions. In the case of iron, it was found that resistivity also increased as density decreased.[3] Resistivity data for these two alloys are shown in Fig. 17. Note that addition of 0.45% phosphorus virtually doubles the resistivity and that it also increases linearly as density decreases. Since the slopes are similar, the effect of the porosity is similar for both alloys.

Aging

It must be stated again and again that, if good properties are to result, the powder used must be pure, pores should be as large as possible, and good sintering practice must be employed. In the work described, all sintering was accom-

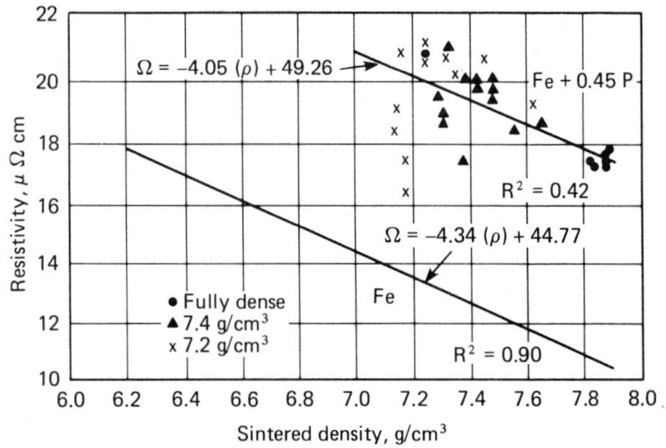

Fig. 17. Predictor equation for resistivity of iron and Fe-0.45P as a function of density

plished using dry hydrogen. Some parts producers use vacuum furnaces which provide properties equivalent to those of parts that are sintered in dry hydrogen.(10) Most parts producers, however, employ dissociated ammonia or atmospheres containing lesser amounts of hydrogen. Whenever these atmospheres are used, the designer must be aware that magnetic properties may be degraded—especially structure-sensitive properties.

A study of the effects of sintering atmosphere on aging characteristics of iron compacts at a density of 7.2 g/cm^3 is currently under way. These compacts were sintered at 2050°F (1120°C). Further study is planned for the iron phosphorus alloys. In the case of iron compacts, aging studies are being conducted on compacts sintered in nitrogen-bearing atmospheres. The aging effects of the nitrogen-base atmospheres on induction at an applied field of 15 oersteds (1190 A/m) are shown in Fig. 18. Although induction is dependent on density, it can be seen that induction is lower when compacts are sintered in pure nitrogen, probably because of nitride formation. Some minor loss in induction also occurs in atmospheres containing less than 20% hydrogen after aging for 1000 hours at 100°C. In general, though, induction is not appreciably

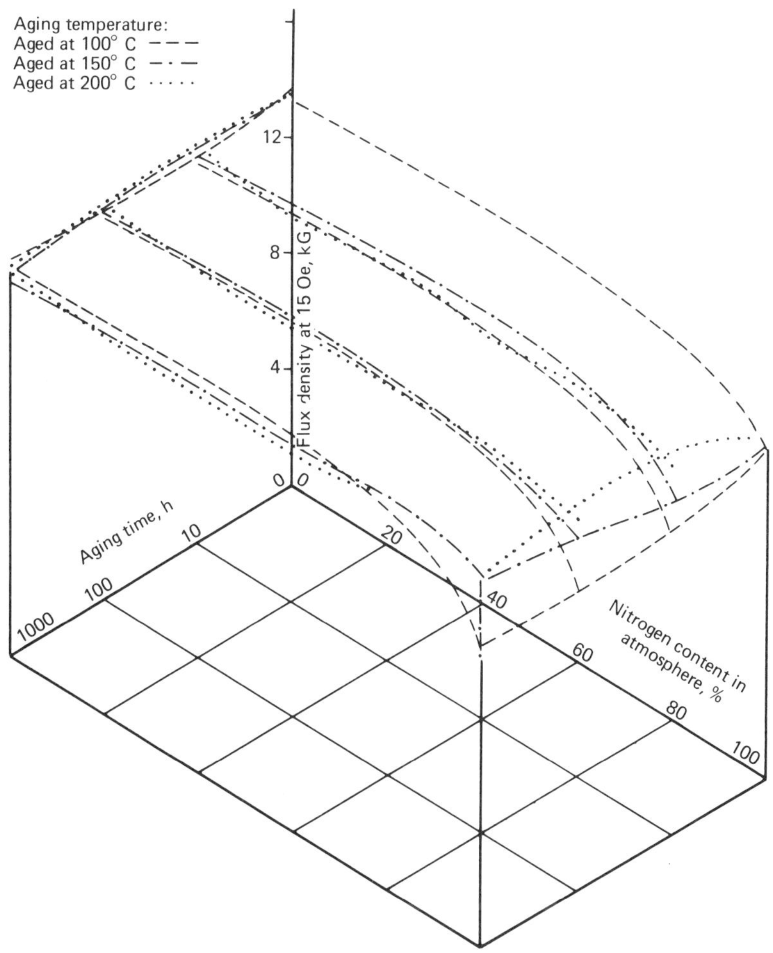

Fig. 18. Effect of aging on induction of iron compacts

lowered if some hydrogen is present in the sintering atmosphere.

Aging also occurs in 7.2 g/cm^3 sintered iron compacts regardless of atmosphere (including hydrogen). The aging characteristics on remanent magnetization of iron in nitrogen-base atmospheres are shown in Fig. 19. Remanent magnetization is lowest if at least 80% hydrogen is present

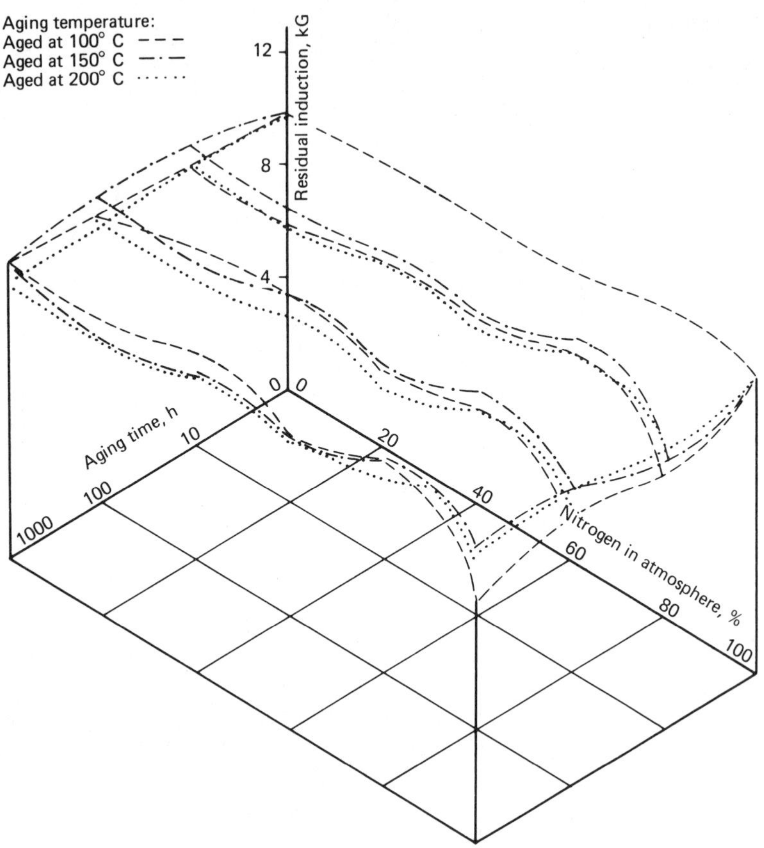

Fig. 19. Effect of aging on residual induction of iron compacts

in the atmosphere, but aging is a problem and causes unpredictable performance with time.

The effect of aging on maximum permeability is shown in Fig. 20. Aging is much more severe when atmospheres containing less than 80% hydrogen are employed; but aging also occurs in compacts sintered in richer hydrogen atmospheres. Compacts sintered in pure nitrogen have the lowest permeability by far. If structure-sensitive properties are required, it is therefore important to avoid lean hydrogen atmospheres if consistent performance is desired.

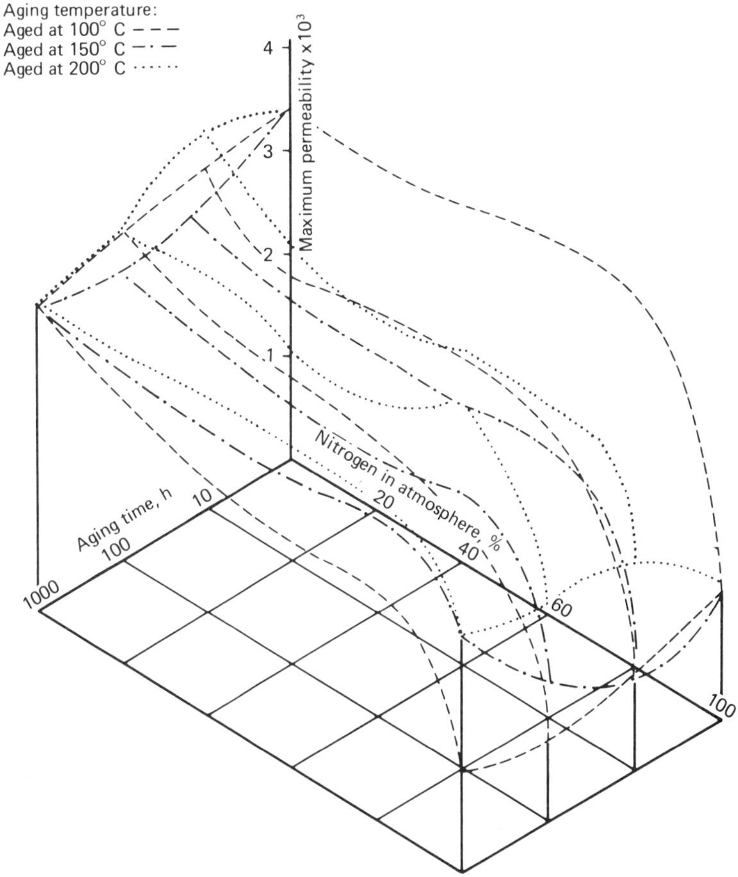

Fig. 20. Effect of aging on maximum permeability of iron compacts

The same results hold true for coercive force (Fig. 21). Compacts sintered in pure nitrogen have extremely high coercive forces that vary dramatically depending on aging conditions. Aging also occurs when iron compacts are sintered in all other atmospheres but most consistent properties are obtained with minimum variation with time and temperature if hydrogen-rich atmospheres are employed for sintering.

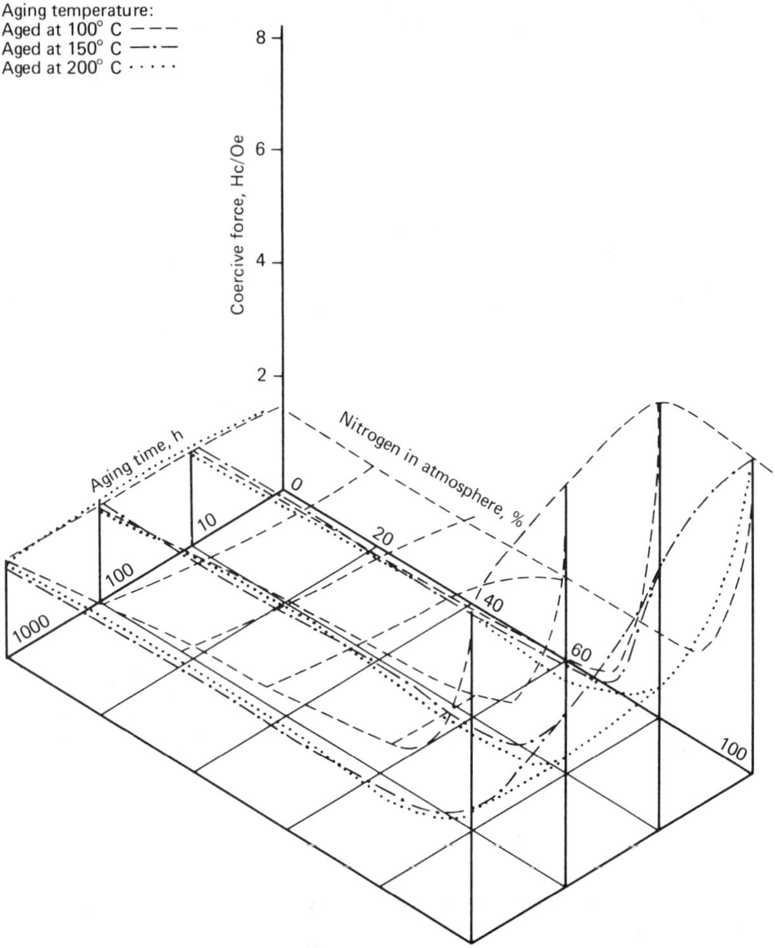

Fig. 21. Effect of aging on coercive force of iron compacts

Not too much work has been done to determine the aging characteristics of phosphorus iron compacts, but from Swedish work early indications are that the 0.45% phosphorus iron alloys do not age, at least if dissociated ammonia is employed as the sintering atmosphere. Höganäs of Sweden has shown this effect on the coercive force[5], as shown in Fig. 22. It can be seen that, as for the previous

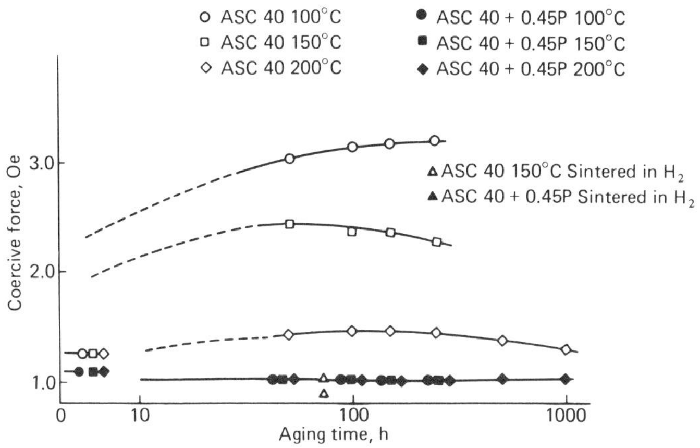

Fig. 22. Effect of aging on ASC 40 and ASC 40 + 0.45%P materials sintered in dissociated ammonia or in hydrogen

data shown, iron compacts age readily in dissociated ammonia; the 0.45% phosphorus iron alloys are invariant. Further work remains to be done to determine the aging characteristics of the phosphorus iron in leaner hydrogen atmospheres.

The above work clearly demonstrates that control of nitrogen in the sintering atmosphere is of utmost importance if iron compacts are to be employed for applications where optimum structure-sensitive properties are required. This is also true of carbon control. Much has been written about the effect of carbon on structure-sensitive properties of wrought materials.[4] Many parts producers employ the same sintering furnaces for sintering both carbon-bearing compacts and compacts for magnetic applications. Many parts producers also change atmospheres within furnaces at will or attempt to sinter compacts for magnetic applications in Endogas or other leaner hydrogen-bearing atmospheres. If this is done there is a strong chance of carbon or nitrogen pickup, which causes severe deterioration of magnetic properties. The effects of these practices are shown by the magnetization curves for iron and 0.45% phosphorus iron compacts in Fig. 23. Magnetization curves

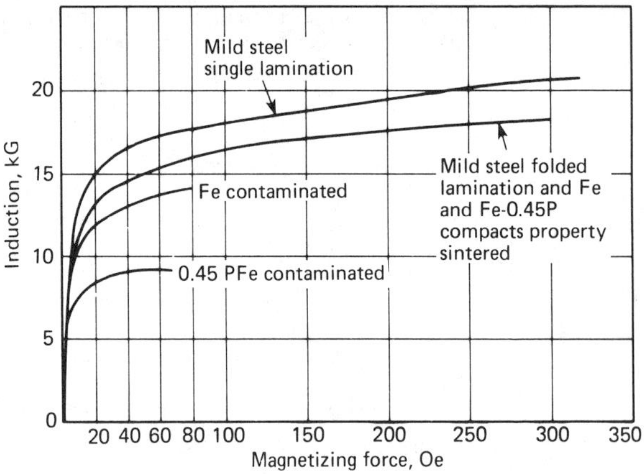

Fig. 23. Comparison of magnetization curves for P/M laminations and mild steel laminations

are shown for several cores that have been sintered in several atmospheres. The upper magnetization curve is for a 1010 AISI steel commonly used for many magnetic applications. The one below is for 0.45% phosphorus iron compacts and iron compacts sintered in hydrogen so that carbon levels are less than 0.01%. Note that the induction is approximately 10% lower because the density is only 7.4 g/cm^3. The other two curves show the drastic lowering of the magnetization curve due to pickup of 0.03% carbon and 0.005% nitrogen because carbon-bearing parts were sintered prior to sintering for magnetic properties.

Not much work has been done in this area, but a small exploratory experiment was made to determine the effect of carbon pickup from contaminated furnaces on phosphorus iron compacts. In this work, magnetic properties were compared for compacts containing 0, 0.45 and 0.8% phosphorus. One set of compacts was sintered in a laboratory furnace where no carbon pickup resulted. These compacts contained less than 0.01% carbon. The other compacts were sintered in a production sintering furnace which was often used for parts containing up to 1% graphite admixed with the powders. The induction levels of these alloys at an applied field of 15 oersteds (1190 A/m) are compared in Fig. 24. The induction of the compacts sintered in the production furnace is slightly lower. The effect of carbon pickup on remanent magnetization is shown in Fig. 25. The remanent magnetization of the compacts that were production sintered was much lower. The effect on permeability is shown in Fig. 26. Note that the permeability of all the compacts that picked up carbon has decreased markedly; however, once again the beneficial effect of the phosphorus addition is clearly seen for this property. The same observation may be made for

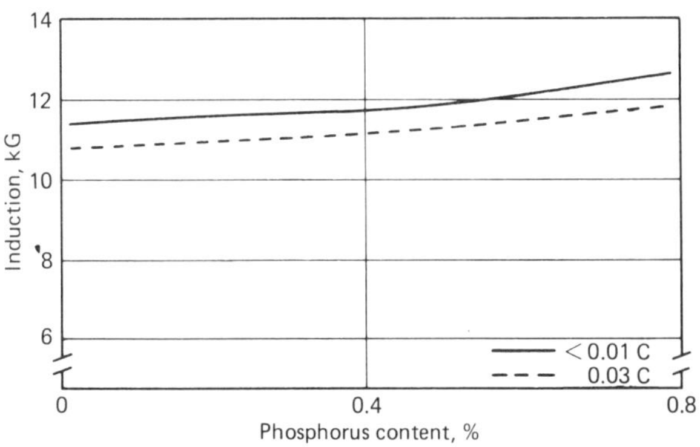

Fig. 24. Effect of carbon pickup on induction

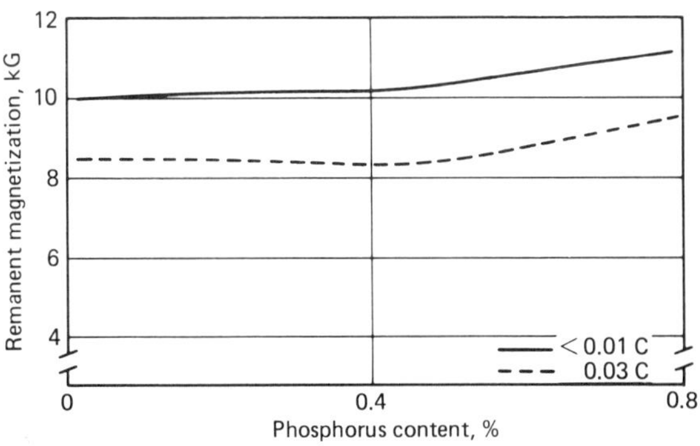

Fig. 25. Effect of carbon pickup on remanent magnetization

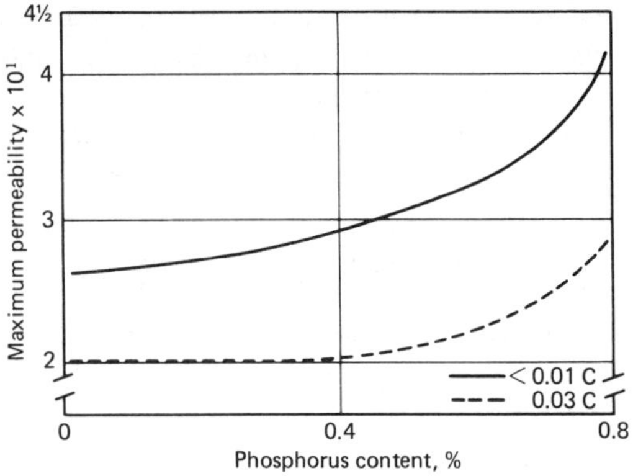

Fig. 26. Effect of carbon pickup on permeability

coercive force (Fig. 27). Although not much work has been done, it is clearly indicated that if a parts manufacturer wishes to fabricate parts in which structural properties are required, the furnace must be maintained so that no contamination occurs during sintering.

Conclusions

It has been shown that there is clearly a strong potential for utilization of phosphorus iron compacts in magnetic

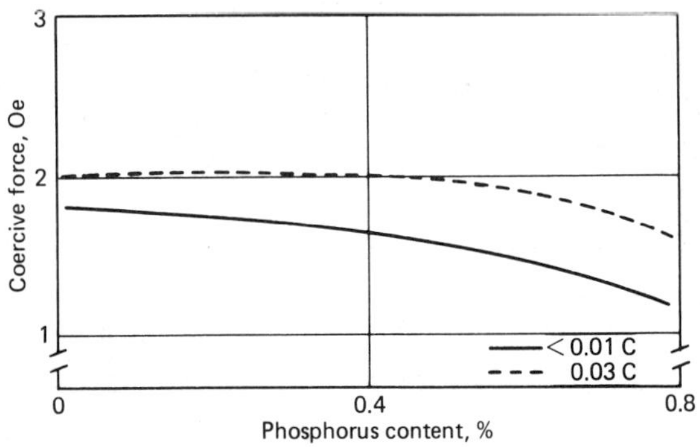

Fig. 27. Effect of carbon pickup on coercive force

applications. These alloys can be made only by P/M techniques, because the hot working required for wrought material renders the alloy hot short. This does not occur in P/M processing, because no hot working is required. If phosphorus iron is fully dense, the saturation induction is only 500 gausses (0.05 teslas) lower. Values continue to decrease as density decreases because less material per volume is available to be magnetized. Remanent magnetization also decreases as density decreases; but it can be further controlled at lower values by using powders and sintering conditions that provide compacts containing large pores. Structure-sensitive properties are a function of grain size. If phosphorus is alloyed with the iron, maximum permeability is significantly increased. The coercive force is also a function of oxygen content. As oxygen levels within the compact are reduced, structure-sensitive properties reach optimum levels. Resistivity is doubled if 0.45% phosphorus is added. All properties may be predicted by simple linear or power equations obtained through regression analysis. Any contamination by either carbon or nitrogen, resulting from carbon pickup or lean hydrogen atmospheres, results in deterioration or variation of structure-sensitive properties as a result of aging. If dissociated ammonia is used for sintering, a 0.45% phosphorus iron does not age.

References

1. G. A. Bolze and J. M. Capus, "Factors Affecting Magnetic Properties of Sintered Iron", in *Modern Developments in Powder Metallurgy*, Vol 11, edited by Pierre Taubenblat, P/M Special Materials and Applications, Metal Powder Industries Federation, Princeton, NJ, 1977, pp 355-370
2. Kenneth H. Moyer, "The Magnetic Properties (DC) of Atomized Iron Powder Cores", in *Modern Developments in Powder Metallurgy*, Vol 11, edited by Pierre Taubenblat, P/M Special Materials and Applications, Metal Powder Industries Federation, Princeton, NJ, 1977, pp 371-384
3. Kenneth H. Moyer, Michael J. McDermott, Mark J. Topolski and David F. Kearney, "Magnetic Properties of Iron Alloys", Magnetic and Electrical P/M Technology and Applications, 1980 International Powder

Metallurgy Conference, Washington, DC, June 25, 1980, Metal Powder Industries Federation, Princeton, NJ

4. Richard M. Bozorth, *Ferromagnetism*, D. Van Nostrand Co., Inc., Princeton, NJ, 1951, pp 251-253
5. Jan Tengzelius and Sten-Ake Kvist, "The Influence of Particle Size and Phosphorus Additions on the Soft Magnetic Properties of Sintered High Purity Atomized Iron Powder", P/M 78, SEMP5, European Symposium on Powder Metallurgy, Stockholm, Sweden, June 4-8, 1978, Jernkontoret-Activity Group 8, Volume II, pp 46-54
6. Kenneth H. Moyer and George W. Mickle, "A Comparison of the Magnetic Properties of Hot Repressed and 7.4 g/cm^3 Iron and 0.45% Phosphorus Iron Cores", SAE Technical Paper 8100239, International Congress and Exposition, Cobo Hall, Detroit, MI, February 23-27, 1981, Society of Automotive Engineers, Inc., Warrendale, PA
7. B. Weglinske and J. Kaczmar, "Effect of Fe_3P Addition on Magnetic Properties and Structure of Sintered Iron", *Powder Metallurgy*, 1980, No. 4, pp 210-216
8. T. Sakai, Y. Suzuki, L. Shimosato, M. Okada and Z. Hermi, "Magnetic Properties of Fe-Si Alloys by Powder Metallurgy", IEEE Transactions on Magnetics, MAG-13, No. 5, September, 1977, pp 1445-1447
9. K. H. Moyer, "P/M Technical Conference", Philadelphia, PA, October 23-26, 1978, Hoeganaes Corporation, Riverton, NJ
10. George Hopple, "Vacuum Sintering of Soft Magnetic Alloys", Senior Thesis, Materials Engineering Department, Drexel University, Philadelphia, PA, 1978

Chapter 8

Aluminum P/M — Properties and Applications

John D. Generous
American Metals Company
Wayne C. Montgomery
Alcoa

Commercial production of precision parts by powder metallurgy (P/M) techniques represents an important new dimension in the fabrication of aluminum alloys.[1] The exceptional properties of aluminum P/M parts–high strength, light weight, corrosion resistance, high thermal and electrical conductivity, and response to a variety of finishing processes–have been utilized in automobiles, appliances, business machines, power tools and many other applications.

Aluminum P/M parts are competitive with many aluminum castings, extrusions and screw machine products that require expensive machining operations. They also compete with P/M parts manufactured from other metal powders in applications where some features of aluminum are needed.

All the physical and mechanical properties of aluminum P/M alloys should be considered by the design engineer in order to fully utilize the capabilities and advantages of aluminum P/M parts.

Material Characteristics[2]

The commercially available aluminum powder alloys are blends of carefully sized atomized aluminum powder mixed with powders of various alloying elements. The main grades, all of which respond to thermal treatment, are essentially comparable to *2xxx* and *6xxx* series wrought aluminum alloys.

The P/M alloy most similar to wrought alloy 2014 is 201AB. It develops high strength and offers moderate corrosion resistance.

Alloy 601AB is similar to wrought alloy 6061. It offers intermediate mechanical properties, good ductility, good corrosion resistance and good finishing characteristics.

Alloy 201AB in the T4 temper (solution heat treated) is the most widely used alloy.

Nominal compositions of P/M premixes are given in Table 1.

Properties and Performance

Aluminum P/M parts can be produced with a range of property values.[3] Tensile strength can vary from 16,000 to 50,000 psi (110 to 345 MPa) depending on composition, density, sintering practices, thermal treatment and repressing operations. Table 2 lists typical properties that can be achieved in four aluminum P/M alloys. These values are affected by variations in compacting, sintering and heat treating practices and should be considered as representative and not minimum properties. Table 3 describes the thermal treatments used for aluminum P/M alloys.

Table 1. Nominal compositions of Alcoa aluminum P/M premixes(a)

	Nominal composition, wt %				
Grade	Copper	Silicon	Magnesium	Lubricant(b)	Aluminum(c)
601AB...	0.25	0.6	1.0	1.5	Rem
602AB...	...	0.4	0.6	1.5	Rem
201AB...	4.4	0.8	0.5	1.5	Rem
202AB...	4.0	...	...	1.5	Rem

(a) Alcoa aluminum P/M premixes are available without lubricant for isostatic pressing or die-wall lubrication. These grades are designated AC rather than AB. (b) Acrowax C, made by Glyco Chemicals, Inc., Greenwich, CT. (c) Grade 1202.

Table 2. Typical properties of Alcoa aluminum P/M alloys

P/M premix	Compacting pressure, tsi	Green density %	Green density g/cm³	Green strength, psi	Sintered density %	Sintered density g/cm³	Dimensional change, in./in.	Thermal condition (temper)	Tensile strength(a), psi	Yield strength(a), psi	Elongation(a), % (in./in.)	Hardness
601AB	7	85	2.29	450	91.1	2.45	-0.019	T1	16,000	7,000	6.0	55-60 HRH
								T4	20,500	14,000	5.0	80-85 HRH
								T6	26,500	25,500	1.0	70-75 HRE
	12	90	2.42	950	93.7	2.52	-0.010	T1	20,100	12,700	5.0	60-65 HRH
								T4	24,900	16,600	5.0	80-85 HRH
								T6	33,600	32,500	2.0	75-80 HRE
	25	95	2.55	1500	96.0	2.58	-0.002	T1	21,000	13,700	6.0	65-70 HRH
								T4	25,600	17,000	6.0	85-90 HRH
								T6	34,500	33,400	2.0	80-85 HRE
602AB	12	90	2.42	950	93.0	2.55	-0.012	T1	17,500	8,500	9.0	55-60 HRH
								T4	17,500	9,000	7.0	65-70 HRH
								T6	26,000	24,500	2.0	55-60 HRE
	25	95	2.55	1500	96.0	2.58	-0.002	T1	19,000	9,000	9.0	55-60 HRH
								T4	19,500	9,500	10.0	70-75 HRH
								T6	27,000	25,000	3.0	65-70 HRE
201AB	8	85	2.36	600	91.0	2.53	-0.018	T1	24,500	21,000	2.0	60-65 HRE
								T4	30,500	26,000	3.0	70-75 HRE
								T6	36,000	36,000	0.0	80-85 HRE
	13	90	2.50	1200	92.9	2.58	-0.007	T1	29,200	24,600	3.0	70-75 HRE
								T4	35,600	29,800	3.5	75-80 HRE
								T6	46,800	46,700	0.5	85-90 HRE
	30	95	2.64	2000	97.0	2.70	-0.003	T1	30,300	26,200	3.0	70-75 HRE
								T4	38,000	31,000	5.0	80-85 HRE
								T6	48,100	47,500	2.0	90-95 HRE
202AB:												
Compacts	13	90	2.49	780	92.4	2.56	-0.011	T1	23,200	10,900	10.0	55-60 HRH
								T4	28,200	17,200	8.0	70-75 HRH
								T6	33,000	21,300	7.3	45-50 HRE
Cold formed parts (19% strain)	13	90	2.49	780	92.4	2.56	-0.011	T2	33,900	31,400	2.3	80 HRE
								T4	34,300	21,500	8.0	70 HRE
								T6	39,800	25,100	8.7	85 HRE
								T8	40,600	36,200	3.0	87 HRE

(a) Tensile properties determined using powder metal flat tension test bar (MPIF Standard 10-63), sintered 15 min at 1150°F (621°C) in nitrogen (average dew point, -45°F or -43°C).

Table 3. Descriptions of various tempers for sintered aluminum P/M alloys

Temper	Description
O	Annealed 1 h at 775°F, furnace cooled at maximum rate of 50°F per h to 500°F or lower
T1	Cooled from sintering temperature to 800°F (601AB and 602AB) or 500°F (201AB) in N_2, air cooled to room temperature
T4	Heat treated 30 min at 970°F (601AB and 602AB) or 940°F (201AB) in air, cold water quenched and aged minimum of 4 days at room temperature
T6	Heat treated 30 min at 970°F (601AB and 602AB) or 940°F (201AB) in air, cold water quenched and aged 18 h at 320°F
T61	Re-pressed, heat treated to T6

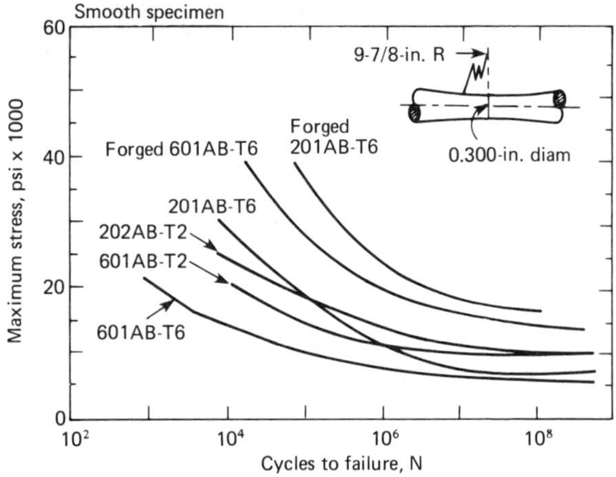

Fig. 1. Typical fatigue behavior of alloys 601AB, 201AB and 202AB

Fatigue is an important design consideration for P/M parts subject to dynamic stresses. Figure 1 shows typical fatigue behavior of specimens of alloys 601AB, 201AB and 202AB in the T2 (as-cold-formed after sintering) and/or T6 tempers.

Corrosion Resistance[3]

Aluminum alloys enjoy widespread use in both structural and nonstructural applications because of their resistance to corrosion. Within the aluminum alloy systems, some compositions are more resistant than others. The Al-Mg-Si alloys (6*xxx* series) exhibit higher resistance to general corrosion than the Al-Cu alloys (2*xxx* series).

In applications requiring high resistance to weathering or in seacoast environments, 601AB is the preferred alloy. Alloy 201AB provides higher strength with some sacrifice in corrosion resistance. The corrosion resistance of aluminum P/M alloys can be improved appreciably through application of chemical conversion coatings or anodizing treatments. Amorphous chromate coatings are especially useful in providing economical protection of parts exposed to saline environments. When, however, exceptional corrosion resistance is required, anodizing treatments are best.

Relative corrosion ratings and descriptions of appearance for two aluminum P/M alloys and two wrought aluminum alloys exposed to 100% relative humidity at 125°F (51°C) for 18 weeks and 5% continuous salt spray for 336 hours are presented in Table 4.

Table 4. Corrosion performance of aluminum P/M and wrought alloys in high humidity and 5% continuous salt spray tests(a)

Alloy	Temper	Relative corrosion rating	Specimen appearance
601AB	O	B	Moderate attack; scattered pits (some large)
	T4	A	Slight attack; scattered small pits
	T6	A	Slight attack; few small pits
201AB	O	D	Heavy attack; many large sites
	T4	C	General attack; scattered large sites
	T6	C	General attack; scattered large sites
6061 (wrought)	T6	A	Slight attack; few small pits
2014 (wrought)	T6	C	General attack; scattered large sites

(a) Average results for 85 and 95% density specimens.

Light Weight[4]

A distinguishing characteristic of P/M aluminum is its lighter weight in comparison with other common P/M materials. Aluminum enjoys a 3-to-1 weight advantage over iron (Fig. 2) and a 3.3-to-1 advantage over copper. This lightness reduces power requirements in drive motors and minimizes vibration and noise in machinery having unbalanced rotary or reciprocating motions.

Conductivity[4]

Another positive benefit of aluminum is its excellent conductivity, both electrical and thermal, in comparison with most other metals. Aluminum P/M parts, therefore,

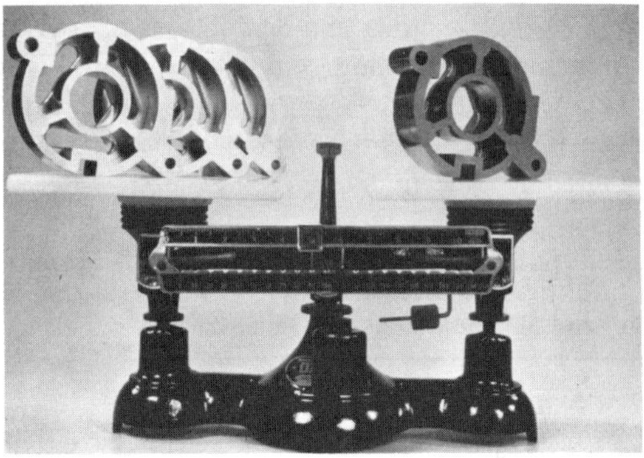

Fig. 2. Aluminum (left) vs iron (right) weight comparison

may be utilized as heat sinks or electrical conductors. Table 5 compares the conductivities of sintered aluminum alloys and various other wrought and P/M materials.

Manufacturing Process

Fabrication of aluminum P/M parts involves the same basic manufacturing operations, equipment and tooling that are employed for other metal powders. Differences in processing parameters are discussed below.

Compacting

Aluminum P/M premixes exhibit excellent compressibility and yield high-density parts at low compaction and ejection pressures[2] (see Fig. 3).

Lower compacting pressures permit the use of smaller, faster presses to produce larger parts or, in some cases, multiple-cavity tooling.

Lower compacting pressures reduce the possibility of damage in fragile tool designs. For parts requiring thin-wall punches (Fig. 4) or small core rods, tool breakage is less likely with aluminum.

Aluminum powder transfers more readily under pressure in the die than do iron powders. It allows molding of more

Table 5. Electrical and thermal conductivity of sintered aluminum alloys and of wrought aluminum, brass, bronze, iron and copper

Material	Thermal condition	Electrical conductivity(a) at 20°C, % IACS	Thermal conductivity(b) at 20°C, CGS units
601AB (2.48 g/cm^3)	T4	38	0.36
	T6	41	0.38
	T61	44	0.41
201AB (2.64 g/cm^3)	T4	32	0.30
	T6	35	0.32
	T61	38	0.36
602AB (2.55 g/cm^3)	T4	44	0.41
	T6	47	0.44
	T61	49	0.45
6061 wrought Al	T4	40	0.37
	T6	43	0.40
Brass (35% Zn)	Hard	27	0.28
	Annealed	27	0.28
Bronze (5% Sn)	Hard	15	0.17
	Annealed	15	0.17
Iron (wrought plate)	Hot rolled	16	0.18
Copper	Hard	97	0.93
	Hot rolled	100	0.93

(a) Determined with FM 103 Magnatester. (b) Converted from electrical conductivity values.

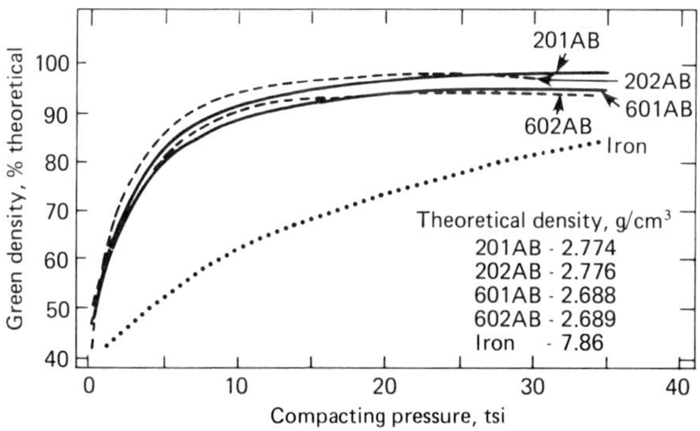

Fig. 3. Compressibility of typical iron P/M powder and aluminum P/M premixes

Fig. 4. Aluminum P/M parts with thin-wall punches

Fig. 5. Aluminum P/M part with molded slope that demonstrates the transferability of aluminum powder during compaction

complex shapes such as sloped and multilevel parts (Fig. 5). It eliminates or reduces the phenomenon of pre-mold, which can be troublesome in compacting certain iron parts.

Sintering[2]

The high strength and good ductility of sintered aluminum P/M parts are directly attributable to the liquid phase that aluminum P/M powders undergo during the sintering process. When aluminum premixes are sintered, liquid phases form as aluminum combines with the added soluble elements copper, silicon and magnesium. These liquid constituents migrate along particle boundaries; they penetrate the aluminum oxide envelope, which is partially fragmented during compaction; and they diffuse into the powder particles. As the particles shift to fill in voids left by the spreading liquid phase, the porosity of the part is reduced. A sufficient oxide network remains during the sintering process to prevent distortion and undesirable shape changes in properly supported parts. The results of liquid-phase sintering include increased intraparticle strength, strong interparticle bonding and homogeneous alloying.

Aluminum P/M parts can be sintered in various types of furnaces and atmospheres. Reproducible dimensions can be achieved with proper attention to compact density, sintering temperature, dew point and atmosphere. Most parts are sintered in a nitrogen atmosphere at 1100°F (593°C) with the dew point maintained at −40°C (−40°F) or better. These conditions develop the highest commercially available properties.

Secondary Operations[5]

Secondary operations performed on aluminum P/M parts are similar to those for most other P/M materials. These include sizing or coining, heat treatment, machining, joining and finishing.

Cold and hot forming are options which can be added when optimum properties are desired.

Sizing, Coining and Cold Forming

Sizing and coining of aluminum P/M parts are re-pressing operations that are performed either to control dimensional tolerances or to increase density, or both. Due to the distortion that may occur during sintering, it is usually necessary to include sizing or coining in the sequence of operations. Exceptions to this rule may occur with parts having relatively low profiles and generous tolerances or with parts that will undergo secondary machining operations.

Typical press tonnages employed in sizing or coining of aluminum P/M parts range from 25 to 40 psi depending on the alloy and the amount of plastic deformation of the part.

Both mechanical properties and surface finish are improved by the coining operation, and detail quite often can be imparted to the surface of a part by coining rather than by molding.

Cold forming is a re-pressing operation in which a significant amount of lateral flow of material occurs. Mechanical-property increases can be realized beyond those obtainable by conventional coining. Upsetting of up to 20% may be involved, and density may be increased to 96 to 99% theoretical.

Heat Treatment

Solution heat treating and aging after sintering will improve mechanical properties and increase ductility of aluminum P/M parts. The same thermal treatment applied to wrought alloys may be used.

Parts should be heated in air at 940 to 1000°F (504 to 538°C) for 30 minutes, then cold-water quenched. Depending on final property requirements, parts can be naturally aged (T4) or artificially aged (T6). The T4 temper provides good strength with maximum ductility. The T6 temper provides maximum strength properties available through heat treating. The additional cost and decrease in ductility of the T6 temper have made the T4 temper the most popular. Heat treatment also improves corrosion resistance and machinability.

Machining

Secondary machining operations performed on an aluminum P/M part may involve any type of metal removal such as drilling, tapping, turning, milling or grinding. Such secondary operations add details of configuration or control tolerances that are not practical in the basic P/M process.

The machinability of P/M aluminum is one of its most attractive features. P/M aluminum provides excellent chip characteristics; chips are small and easily broken, with none of the long stringer buildup that occurs with some wrought aluminum alloys (Fig. 6). This means good tool life, increased speeds and feeds and hence higher production rates.

For optimum machining, parts should be given a T4 or T6 thermal treatment. High speeds and good lubricant flow to carry away chips produce the best results. Sintered carbide inserts are preferred since a sharp cutting edge will prevent smearing. Lard oils, water-soluble oils (3 to 5% in water) and chemical emulsions are satisfactory lubricants.

Fig. 6. Machining chips from a wrought aluminum alloy (left) and from a P/M aluminum alloy (right)

Drilling, tapping and milling can be performed by following the procedures used for wrought aluminum. Good lubricant flow, sharp tools, high speeds and rapid heat removal are the important parameters. Figure 7 illustrates machining characteristics.

Finishing[6]

Sintered aluminum parts can be given a variety of surface finishes. Many of the decorative and protective treatments currently employed for wrought and cast aluminum alloys can also be applied to aluminum P/M parts. These include chemical cleaning, mechanical finishing and etching to achieve textures; anodizing for increased resistance to corrosion, wear and abrasion; coloring for decorative or functional purposes; electroplating; and painting. Some

Fig. 7. Aluminum P/M blank (left) and finished machined part (right)

Fig. 8. Assorted aluminum P/M parts in a variety of finishes

chemical cleaning or etching solutions may be too aggressive for parts of less than 90% density; however, preselection of suitable solutions can be made. Alumilite* and hard-type anodic coatings can be applied in many applications. Thus, the designer has a wide choice of decorative and functional finishes not possible with other P/M materials.

Figure 8 illustrates the variety of finishes that can be applied to aluminum P/M parts.

Joining

Common methods employed for joining of parts include press fitting, staking, adhesive bonding and joining with threaded fasteners. Joining during sintering also works except where the shape of the assembled parts restricts the ability to properly size the structure.

Joining of aluminum P/M parts by copper infiltration is not possible, because the melting point of copper is higher than the sintering temperature of aluminum. Normal welding and brazing techniques are unsatisfactory for joining of aluminum P/M parts. Soldering can be used if the strength of such a bond is adequate.

Where part design allows, press fitting and staking are practical and economical methods of joining two or more

*Trade name of Aluminum Company of America

parts together. These operations also may be used to join non-P/M components to sintered aluminum parts. Dual-flange pulleys are common examples of stampings staked or pressed onto aluminum P/M pulley hubs (Fig. 9).

Prior to joining, P/M parts should be solution heat treated to the T4 temper to ensure maximum ductility and strength. Interference fits of 0.003 to 0.005 in. are recommended to ensure secure components (Fig. 10). Adequate radii and leads should be incorporated in the design to facilitate pressing and to prevent shearing. Keys, D-shape bores and bosses can be utilized where necessary for maximum torque requirements.

Sintered aluminum parts can be successfully fastened by adhesive bonding. Adhesive manufacturers should be consulted for recommendations on individual applications.

Use of threaded fasteners is a more conventional method of joining multiple parts. Although usually not as eco-

Fig. 9. Aluminum P/M dual-flange pulleys

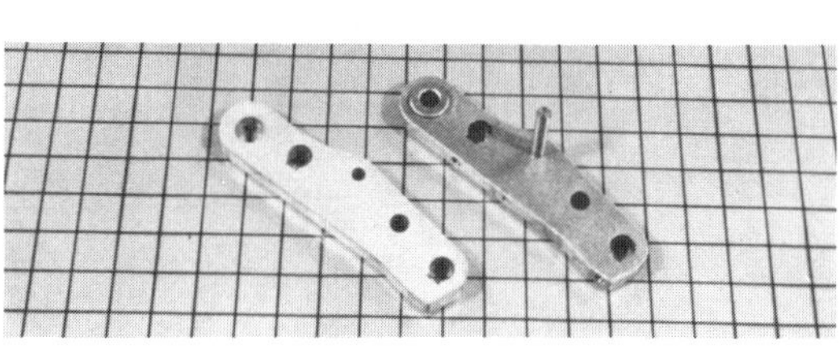

Fig. 10. Aluminum P/M part with press fitted bearing and screw machine pin

nomical as some other methods, joining with threaded fasteners produces secure components and provides for disassembly. Excellent thread characteristics can be obtained in aluminum P/M parts above the 90% density level. The ductility of parts in the upper density range also is sufficient for self-tapping fasteners.

Applications

Aluminum P/M parts are being used today in a wide variety of applications. Parts vary in size and weight from 1/2 in. in diameter and a few grams to about 4 in. in diameter and several hundred grams.[7] The following review of experimental and production parts illustrates their variety and use in various markets. These case histories should be studied by designers who may need the inherent advantages of P/M aluminum.

Business Machines

The business-machine market currently uses the greatest variety of aluminum P/M parts. Figure 11 illustrates a cross section of the types used.

Figure 12 shows drive pulleys used for portable tabletop copying machines. P/M aluminum provides good machin-

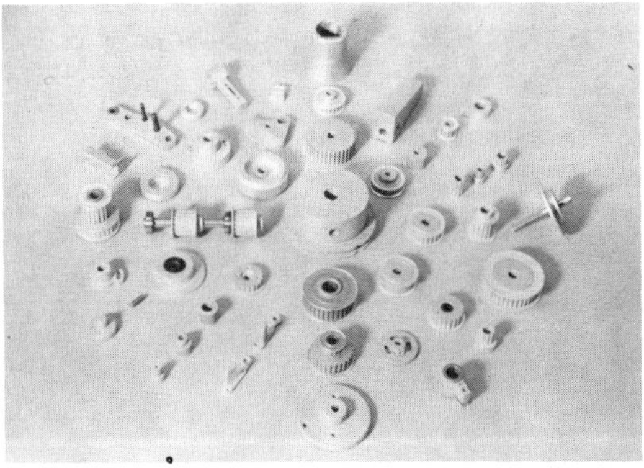

Fig. 11. Assorted aluminum P/M parts for business machines

ability, corrosion resistance and light weight. One part is drilled to produce a second version, to achieve savings in tool cost.

Figure 13 shows a registration stop that functions as a cam stop actuator in a business copying machine. Light weight lowered the inertia for quieter machine operation. Tolerances are critical, with the dimension from the cam stop to the center of the ID held within 0.001 in. Corrosion resistance is another advantage.

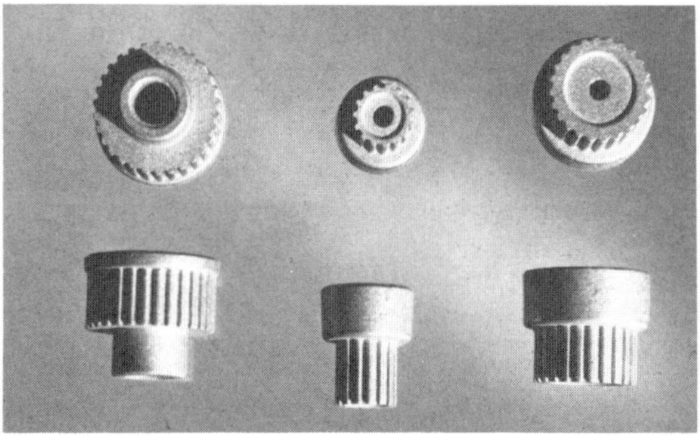

Fig. 12. Aluminum P/M drive pulleys for portable tabletop copying machines

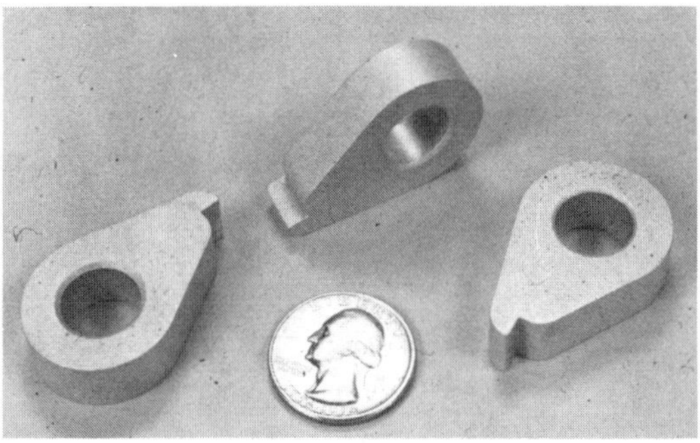

Fig. 13. Aluminum P/M registration stop used as a cam stop actuator in a business copying machine

Figure 14 shows a bearing housing for a computer. Being of aluminum, it lends itself to press fitting of the bearing without distortion. This part was originally machined from stainless steel bar stock. Switching to P/M aluminum resulted in a cost reduction of 40 to 50%.

Figure 15 shows a copying machine hub. Weighing approximately 550 grams, this "hub" was used inside a color copying machine. A larger nylon gear is injection molded onto the flange. Aluminum was preferred since low weight and excellent machinability were prime requirements. Low weight was necessary for quick acceleration and deceleration. Good machinability allowed drilling and tapping of five holes, the special flange contour incorporated in the secondary operations, and machining of the ID.

Fig. 14. Aluminum P/M bearing housing for a computer

Fig. 15. Aluminum P/M hub for a color copying machine

Figure 16 shows a bearing holder that replaced an aluminum extrusion in a copying machine.

The excellent machinability of P/M aluminum made the pad wedge in Fig. 17, which required deep hole drilling, more cost effective than a ferrous P/M part. The unit on the left has been hard coat anodized for wear resistance.

Automotive Components

As the automotive industry continues to strive for weight reduction and improved corrosion resistance, aluminum P/M parts become more and more attractive for use in this market (see Fig. 18).

Shock-Absorber Pistons. Aluminum P/M pistons for automotive shock absorbers, shown in Fig. 19, offered 15%

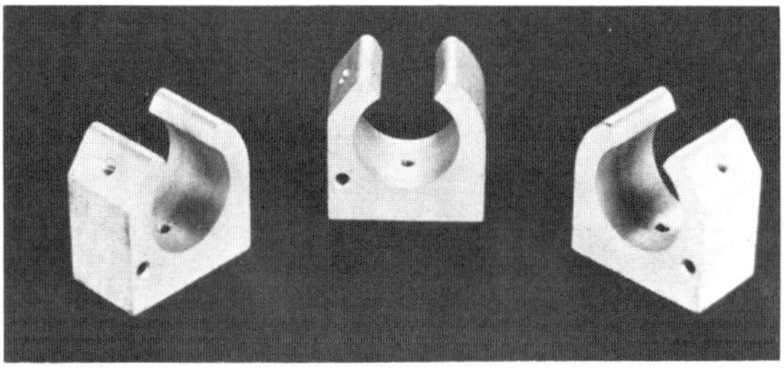

Fig. 16. Aluminum P/M bearing holder that replaced an aluminum extrusion in a copying machine

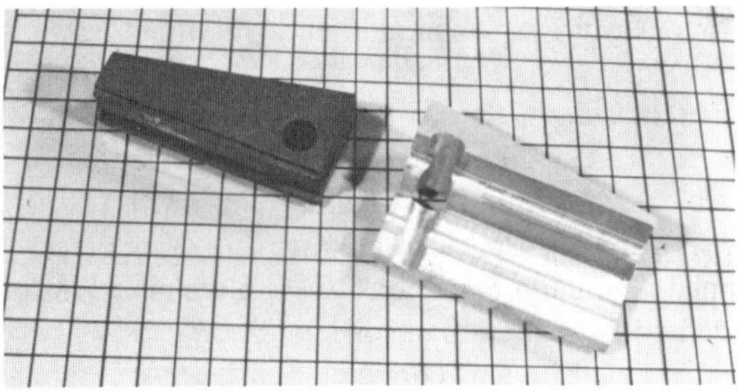

Fig. 17. Aluminum P/M pad wedge

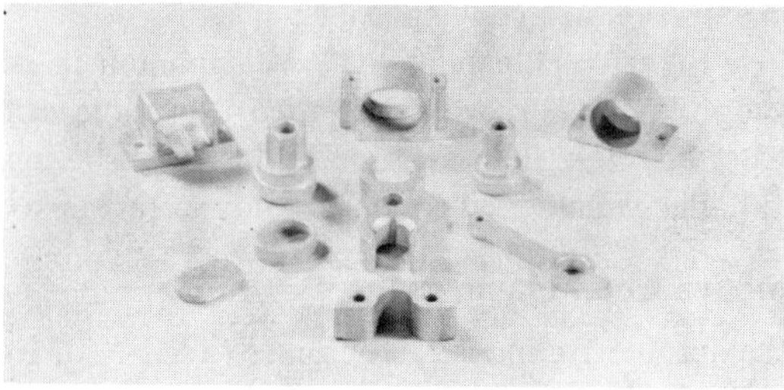

Fig. 18. Assorted aluminum P/M automotive parts

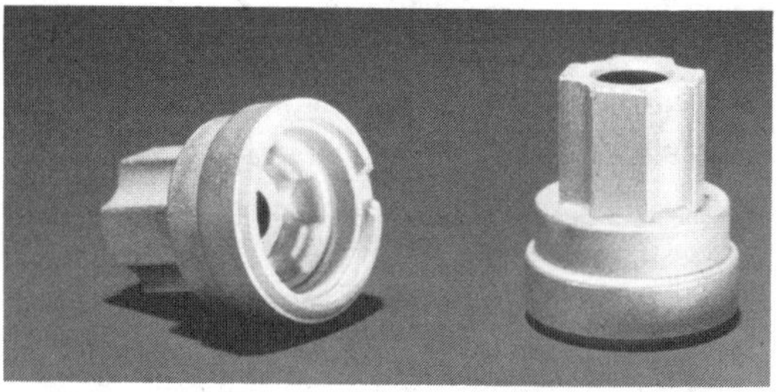

Fig. 19. Aluminum P/M pistons for automotive shock absorbers

more strength than their iron P/M counterparts. Better machinability allowed increased production rates in drilling and tapping. Inherent light weight reduced shipping and handling costs significantly. Good formability provided better assembly staking characteristics. The parts performed extremely well in a fleet of Chicago taxicabs in a 50,000-mile "temperature resistance and durability field test".

Sensor Housings. Sensor housings used in an antiskid device for trucks are shown in Fig. 20. Close tolerances, corrosion resistance and finishability along with tensile and crush strength are the main requirements. Compared with machined aluminum extrusions, these aluminum P/M parts provided a 50% reduction in cost.

The light weight and corrosion resistance of aluminum were required for the automotive air-conditioning part shown in Fig. 21. Redesigning from a machined extrusion to an aluminum P/M part reduced manufacturing costs.

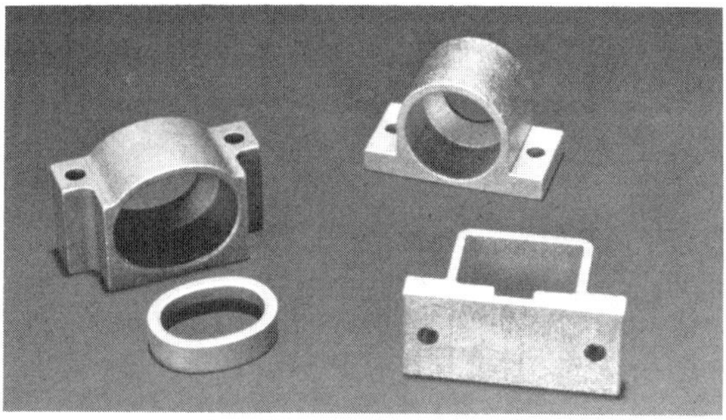

Fig. 20. Aluminum P/M sensor housings for a truck antiskid device

Fig. 21. Aluminum P/M part for an automotive air conditioner

Appliances

Figures 22 and 23 show sewing-machine parts that were used in reciprocating-motion applications where light weight reduces vibration and noise level during movement. Tolerances are critical on the center distance of the two ID's. Parts are heat treated to improve strength.

Figure 24 shows a thermostat control gear that is part of a control mechanism for a countertop oven. The critical area is the half tooth on both ends of the cut gear section. Due to the delicacy of this part, aluminum P/M was the preferred material because of its superior green strength and good corrosion resistance.

Power Tools

Figure 25 shows a saw blade drive holder that was produced originally by machining a steel bar. Production of

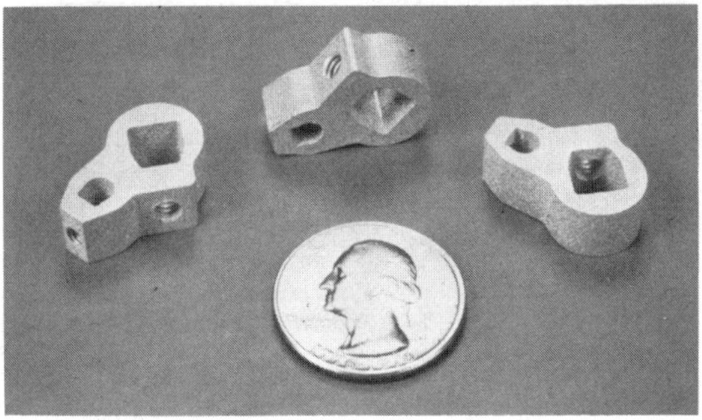

Fig. 22. Aluminum P/M sewing-machine parts

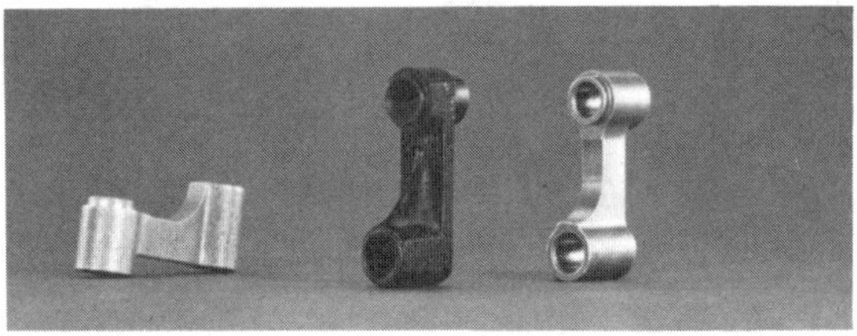

Fig. 23. Aluminum P/M sewing-machine parts

Fig. 24. Aluminum P/M thermostat control gear for a countertop oven

this part from aluminum P/M provided a two-thirds reduction in weight, permitting quicker acceleration and deceleration with less vibration. P/M eliminated troublesome scrap inherent in the machining process, and aluminum offered light weight and good corrosion resistance.

Fig. 25. Aluminum P/M saw blade drive holder that replaced a part made by machining from steel bar

Miscellaneous Applications

Figure 26 shows a sidewinder spacer for a textile machine. Converting from stamping to P/M offered an appreciable cost reduction and closer dimensional control.

The thermal conductivity of aluminum allowed the heat sink shown in Fig. 27 to stay cool during operation.

Figure 28 shows an aluminum P/M part used in a refrigeration control valve.

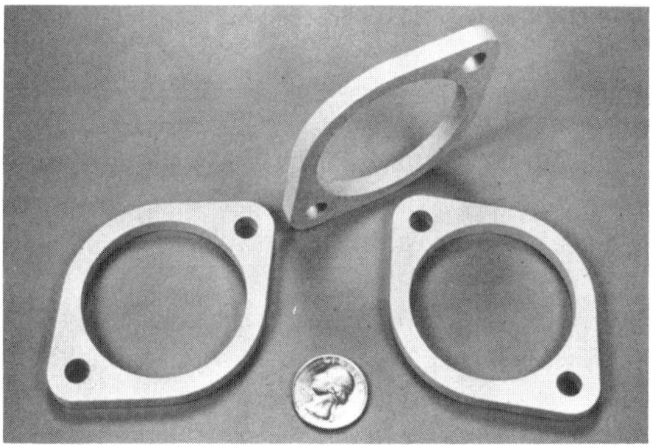

Fig. 26. Aluminum P/M textile-machine sidewinder spacer that replaced a stamping

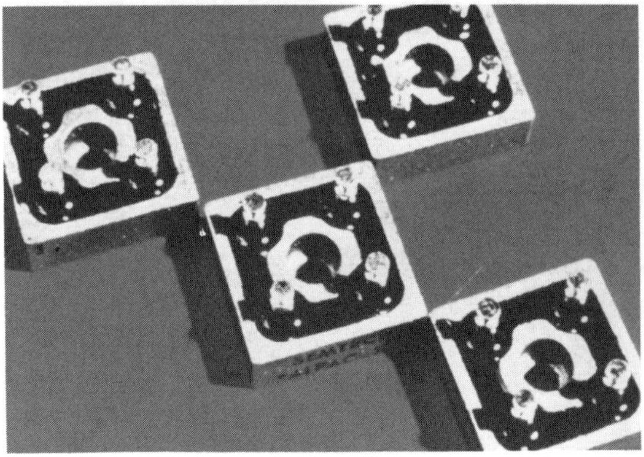

Fig. 27. Aluminum P/M heat sink

Fig. 28. Aluminum P/M part for a refrigeration control valve

Figure 29 shows an aluminum P/M sparkproof connector which proved to be more cost effective than a screw machine part and provided better tolerances on the diametral taper.

Figure 30 shows structural spacers in various OD/ID sizes and thicknesses (over 3/8 in.) that offer a marked cost reduction over conventional stamped spacers. Scrap loss is practically nil. Also, stamping becomes more difficult and less precise as thickness increases. These spacers, used on structural electric transmission towers and other similar applications, offer excellent corrosion resistance. Cost

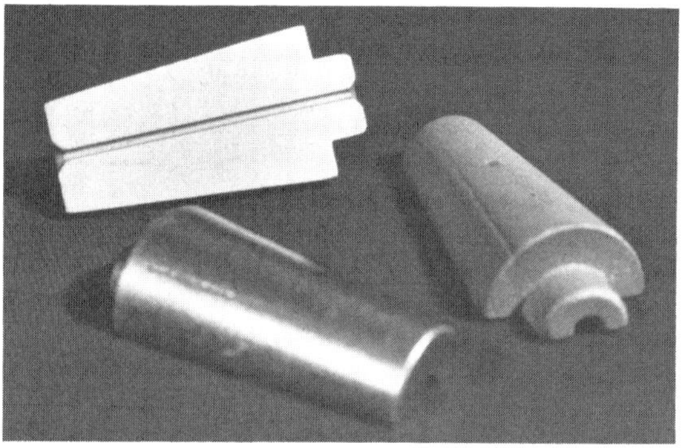

Fig. 29. Aluminum P/M sparkproof connector that replaced a screw machine part

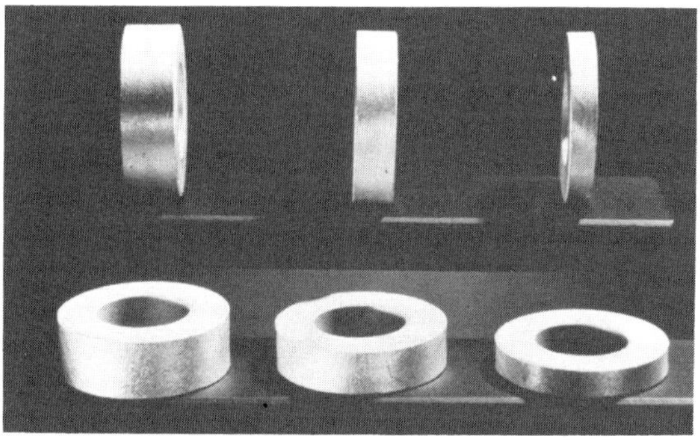

Fig. 30. Aluminum P/M structural spacers that replaced conventional stamped spacers

savings increase significantly with increases in thickness. A 75% cost reduction was realized with a 9/16-in.-thick spacer.

Summary

Aluminum alloys represent an important added dimension to the parts user. Attractive mechanical and physical properties provide the engineer with increased flexibility in material selection and design. The unique characteristics of aluminum will continue to expand the markets and applications for aluminum P/M parts.

References

1. J. H. Dudas and W. A. Dean, "The Production of Precision Aluminum P/M Parts", *International Journal of Powder Metallurgy*, Vol 5, April 1969
2. *Alcoa Aluminum Powders in Powder Metallurgy*, Aluminum Company of America, PAP922, June 1982
3. J. H. Dudas and K. J. Brondyke, "Aluminum P/M Parts—Their Properties and Performance", Technical Paper No. 70041, Society of Automotive Engineers, New York, F38-12903, 1970
4. *The Properties and Performance of Aluminum P/M Parts*, Aluminum Company of America, F38-13347, 1973
5. J. D. Generous, "Secondary Operations for Aluminum P/M", presented at the Society of Automotive Engineers Congress, Detroit, Feb 27, 1979
6. P. Brasko and E. C. Patterson, "Finishing Characteristics of Aluminum P/M Products", Aluminum Company of America, 1970, F38-12856
7. J. D. Draper, M. K. Gala, and C. B. Thompson, "Aluminum P/M Parts Fabrication", presented at the 1976 International P/M Conference, F38-13857

Chapter 9

Wrought Aluminum P/M Alloys

Robert H. Graham
Alcoa

A new area of high technology based on rapid solidification of atomized powder is emerging in the P/M field. This technology employs the atomization process to produce alloyed powders for eventual use as wrought products with engineering characteristics superior to those of aluminum alloys produced by conventional ingot casting procedures. Except for the use of atomized powder as the starting material, this technology is not associated with conventional P/M press-and-sinter processing.

Two wrought P/M alloys currently being commercialized by Alcoa, 7090 and 7091, are zinc-base alloys with additions of cobalt in quantities generally larger than those possible in alloys produced by ingot metallurgy.

Wrought P/M aluminum alloys are produced by a process developed by Alcoa which permits the use of new compositions and the development of finer grain structures, more homogeneous distributions of alloying elements, and optimum combinations of strength, toughness and resistance to corrosion. These alloys are particularly suitable for aerospace parts subject to extraordinary corrosion or mechanical stress. The first two wrought P/M alloys being produced in commercial quantities offer a well-balanced combination of:

- High strength
- Excellent resistance to exfoliation and stress-corrosion cracking

- Combinations of fracture toughness and yield strength that are better than those of most alloys produced by ingot metallurgy (I/M)
- Notched fatigue strengths higher at long life than those of the best I/M alloys.

Typical Tensile Properties

The longitudinal yield strength of extruded P/M alloy 7090-T7E71 (see Table 1) is about 15% higher than that of alloy 7050-T76511, and the yield strength of P/M alloy 7091-T7E69 (Table 1) is 7% higher than that of 7050-T76511.

The longitudinal yield strength of die-forged 7090-T7E71 (see Table 2) is about 18% higher than that of forged 7175-T736, and that of 7091-T7E69 (Table 2) is about 8% higher than that of 7175-T736.

Corrosion Resistance

Resistance to Exfoliation. The EXCO test has become the standard measurement of resistance to exfoliation corrosion by I/M 7*xxx* alloys in the various T7 tempers. Recent laboratory results indicate that the EXCO test procedures may not be applicable to P/M alloys. In long-term natural-environment exposure tests, P/M alloys in the

Table 1. Typical tensile properties of extruded shapes(a) of P/M alloys 7090-T7E71, 7091-T7E69 and 7091-T7E70

Alloy	Tensile strength MPa	Tensile strength ksi	Yield strength MPa	Yield strength ksi	Elongation, %-4D
7090-T7E71:					
L	627	91	586	85	10
LT	593	86	538	78	6
7091-T7E69:					
L	593	86	545	79	12
LT	552	80	510	74	8
7091-T7E70:					
L	545	79	496	72	14
LT	517	75	462	67	10

(a) Thickness, 6.4 to 38.1 mm (0.25 to 1.5 in.)

Table 2. Typical tensile properties of die forgings(a) of P/M alloys 7090-T7E71 and 7091-T7E69

Alloy	Tensile strength MPa	Tensile strength ksi	Yield strength MPa	Yield strength ksi	Elongation, %-4D
7090-T7E71:					
L	614	89	579	84	10
T	579	84	545	79	4
7091-T7E69:					
L	579	84	531	77	13
T	545	79	496	72	9

(a) Thickness, ≤76.2 mm (3.000 in.)

T76 and peak strength T6 tempers show no evidence of exfoliation during exposure over four years to the seacoast atmosphere.

Excellent Resistance to Stress-Corrosion Cracking at Higher Strengths. The P/M alloys combine considerably higher strength with resistance to stress-corrosion cracking which is equal to that of alloy 7075-T73, the I/M alloy considered to be the standard for excellence in these important properties.

Short-transverse specimens taken across the highly directional parting planes of web-flange die forgings were exposed for five years to an industrial atmosphere at a constant stress of 241 MPa (35 ksi). Survival rates of these specimens (vertical axis) are plotted against short-transverse yield strength in Fig. 1. P/M alloys were superior to I/M alloys by 69 to 103 MPa (10 to 15 ksi).

High Resistance to Atmospheric Conditions. Specimens of 7090 and 7091 forgings, extrusions and rolled plate were exposed to the industrial atmosphere at Alcoa Technical Center near Pittsburgh, Pennsylvania, and to the seacoast environment near Point Judith, Rhode Island. Corrosion of all specimens was uniform and nondirectional. Preliminary results from continuing tests indicate that coatings and other surface treatments used commonly in the aerospace industry protect P/M and I/M alloys equally well.

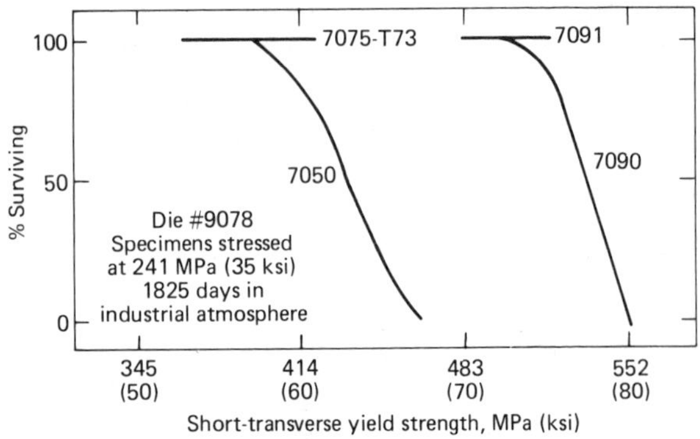

Fig. 1. SCC strength capabilities of thin web-flange die forgings of P/M alloys 7090 and 7091 and I/M alloys 7050 and 7075

Fracture Toughness/Strength Relationships

Initial testing of the strength/toughness relationships of the new P/M alloys indicates their superiority to most of the *7xxx* series I/M alloys. The relationships between yield strength and fracture toughness for extruded shapes 6.4 to 38.1 mm (0.25 to 1.50 in.) thick in their L-T orientation are shown in Fig. 2. Points approaching the upper right-hand corner of the graph are preferred.

Notched Fatigue Strength

The notched axial fatigue strengths of alloys 7090 and 7091 are 35 to 40% higher than those of alloys 7050, 7075 and 2024 (an I/M alloy often selected for its resistance to fatigue) at one million or more cycles (see Fig. 3 and 4).

Nominal Compositions

Wrought P/M alloys 7090 and 7091—previously known as MA67 and MA87, and then CT90 and CT91—are in the *7xxx* series of alloys; zinc is their main alloying constituent (see Table 3). Addition of cobalt in quantities

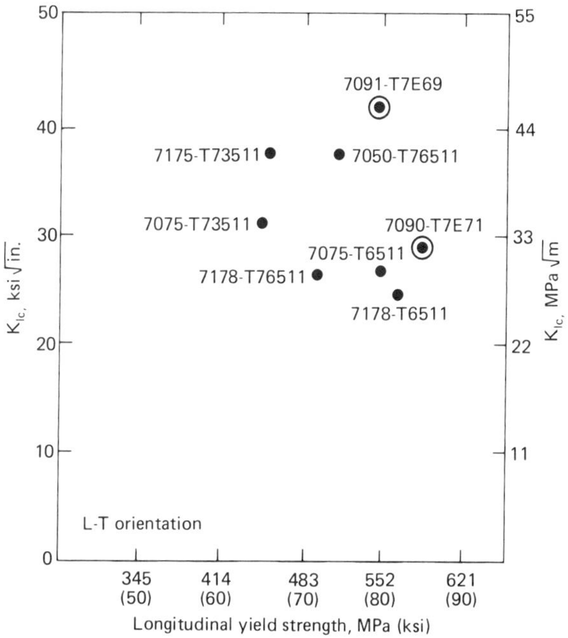

Fig. 2. Fracture toughness vs yield strength for high-strength aluminum alloy extruded shapes 6.4 to 38.1 mm (0.25 to 1.50 in.) thick

generally larger than those possible in alloys produced by ingot metallurgy—1.5% in 7090 and 0.4% in 7091—refines the grain structure. Rapid solidification during atomization ensures homogeneous distribution of alloying elements.

Grain Structure

The photomicrographs of I/M alloy 7050 and P/M alloy 7091 shown in Fig. 5, both at a magnification of 500X, demonstrate the finer grain structure of 7091 and its more homogeneous dispersal of alloying elements. The finer, more homogeneous structure contributes significantly to the superior resistance to stress-corrosion cracking and exfoliation demonstrated by P/M alloys.

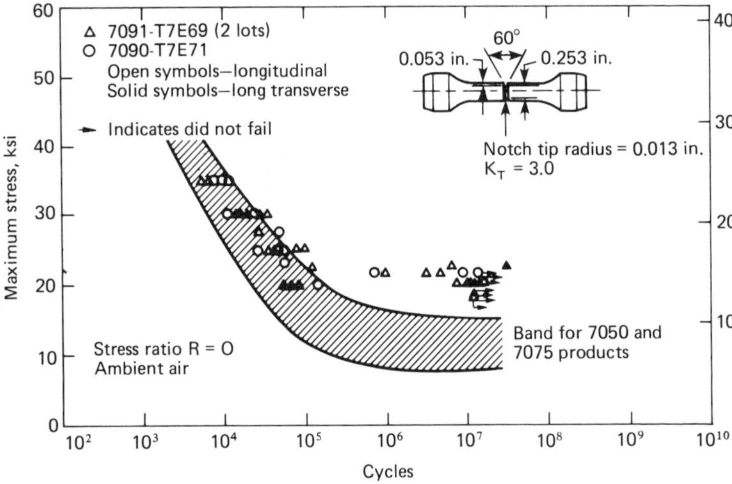

Fig. 3. Comparisons of notched axial fatigue strength for P/M alloy 7090 and 7091 extrusions vs I/M alloy 7050 and 7075 products

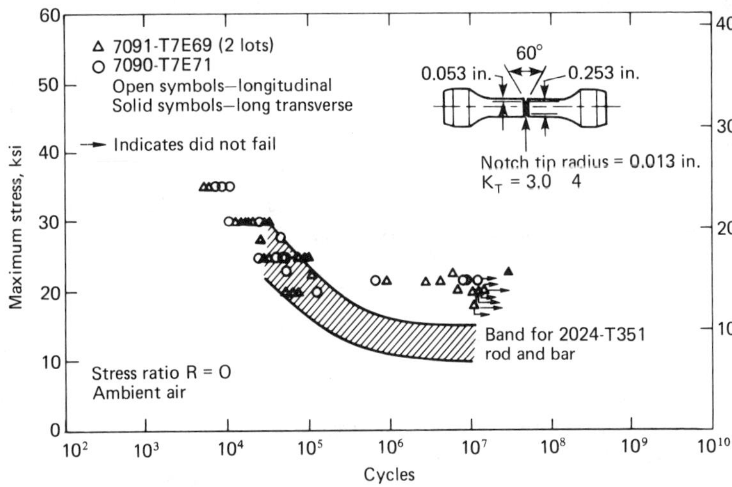

Fig. 4. Comparison of notched axial fatigue strength for P/M alloy 7090 and 7091 extrusions vs I/M alloy 2024-T351 rod and bar

Table 3. Nominal compositions of high-strength aluminum P/M and I/M alloys

Alloy	Nominal composition, wt %						
	Cu	Mg	Cr	Zn	Co	O	Al
P/M Alloys							
7090	1.0	2.5	...	8.0	1.5	0.35	Rem
7091	1.5	2.5	...	6.5	0.4	0.35	Rem
I/M Alloys							
7050	2.3	2.25	...	6.2	...	...	Rem
7075	1.6	2.5	0.23	5.6	...	...	Rem
7175	1.6	2.5	0.23	5.6	...	...	Rem
7178	2.0	2.8	0.26	6.8	...	...	Rem

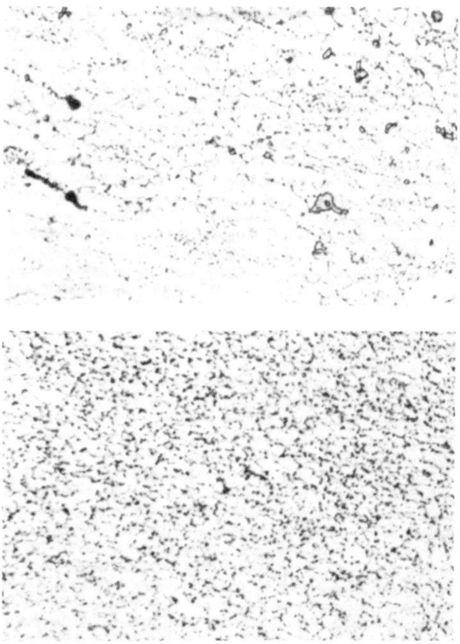

Note the finer grain structure and more homogeneous dispersal of alloying elements exhibited by P/M alloy 7091. Magnification (both), 500X.

Fig. 5. Photomicrographs of I/M alloy 7050 (top) and P/M alloy 7091 (bottom)

Table 4. Physical properties of aluminum P/M alloys 7090 an

Alloy	Solidus temperature (approx) °C	°F	Specific heat(a), cal/g/°C	Thermal conductivity(a), cal/s·°C/cm
P/M Alloys				
7090	548	1018	0.203	0.33
7091	543	1010	0.206	0.36
I/M Alloys				
7050	...	...	...	...
7075	...	...	...	...
7178	...	...	...	...

(a) Calculated values. (b) At 20 to 100°C (68 to 212°F). (c) Valu of tension and compression moduli.

Physical Properties

The physical properties of P/M alloys 7090 and 7091 (and values of density and modules of elasticity for I/M alloys 7050, 7075 and 7178) are presented in Table 4.

Manufacture of P/M Alloys

Wrought P/M aluminum alloys 7090 and 7091 are being produced in billet form at Alcoa's facility near Pittsburgh, Pennsylvania. In a typical production sequence:

1 Molten prealloyed aluminum is atomized and solidified rapidly. Rapid solidification refines the dendritic structure and reduces the size of constituent particles (see Fig. 6).
2 The powder particles are isostatically compacted and encapsulated in an aluminum canister (see Fig. 7).
3 The compact is heated and gases are evacuated from the canister (Fig. 8).
4 The heated canister and compact are pressed into a billet with 100% density (Fig. 9).

7091 and aluminum I/M alloys 7050, 7075 and 7178

Coefficient of expansion(a)(b)		Density		Modulus of elasticity(e)	
n/m/°C	in./in./°F	g/cm^3	lb/in.^{3}	MPa × 10^3	ksi × 10^3
3.8 × 0^{-6}(c)	13.2 × 10^{-6}(c)	2.850	0.103	73.8	10.7
3.7 × 0^{-6}(d)	13.1 × 10^{-6}(d)	2.823	0.102	72.4	10.5
...	...	2.823	0.102	71.7	10.4
...	...	2.796	0.101	71.7	10.4
...	...	2.823	0.102	71.7	10.4

e for T7E71 temper. (d) Values are for T7E69 temper. (e) Average

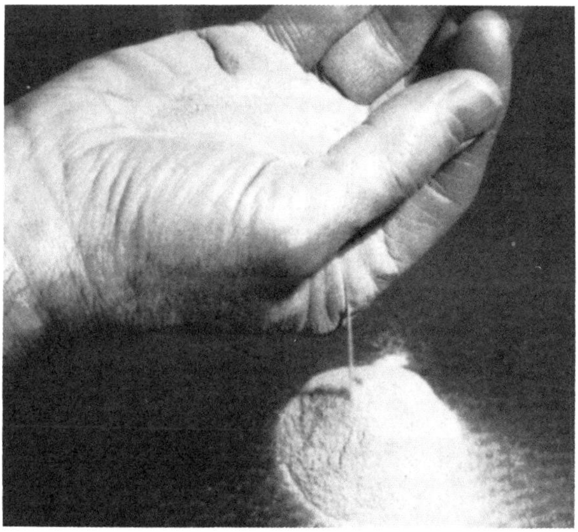

Fig. 6. Aluminum alloy 7090 atomized powder. Average particle diameter, 15-20 μm

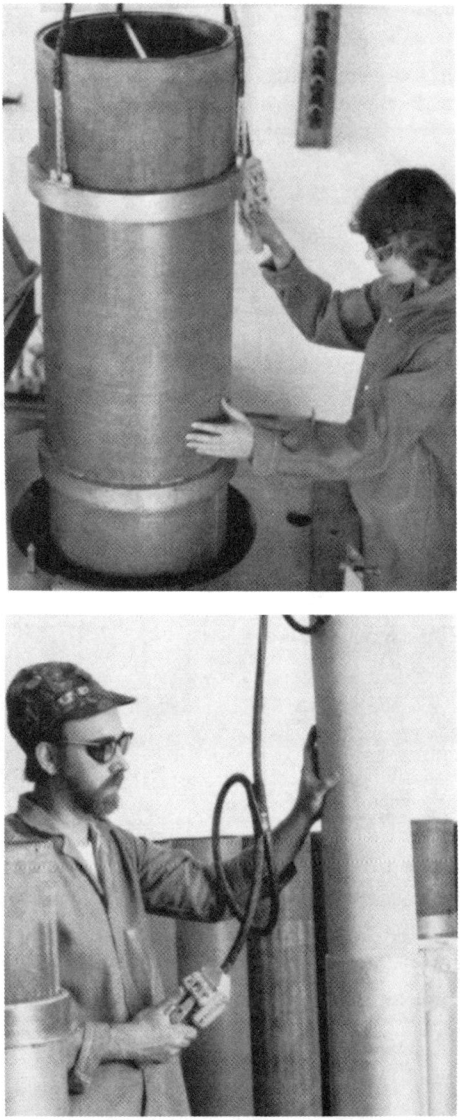

Fig. 7. (Top) Atomized powder and container being lowered into cold isostatic press. Compaction to approx 70% maximum density occurs. (Bottom) Resulting green compact is encapsulated in an aluminum canister for vacuum preheating.

Fig. 8. Encapsulated green compact positioned in furnace for vacuum preheating

Fig. 9. Preheated and degassed green compact readied for final consolidation to 100% density

5 The canister is machined from the surface of the billet (Fig. 10).
6 The billet is ready for subsequent fabrication into parts (see Fig. 11) using a conventional technique such as forging or extrusion. Development of larger billets will permit fabrication of sheet and plate products.

Fig. 10. Aluminum canister being machined off P/M billet surface in preparation for conventional wrought processing

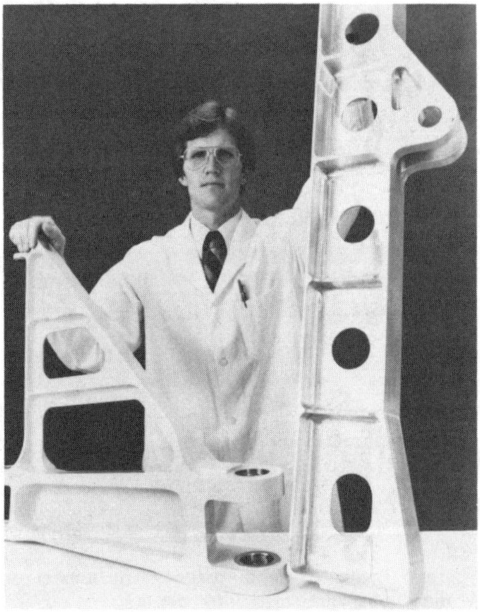

Fig. 11. P/M alloy 7090 forgings used in the main landing-gear assemblies of Boeing's 757 airliner effect a 15% weight savings over previous parts made with conventional aerospace aluminum alloys.

Chapter 10

Powder Metallurgy Parts for Aerospace Applications

J. H. Moll
V. C. Petersen
E. J. Dulis
Crucible Research Center
Colt Industries Inc.

A need exists throughout the aerospace industry to reduce the material, energy and labor required to produce engine and airframe hardware. Typically, the buy-to-fly ratio of these components is high, often reaching 20:1 and higher. This results in high costs for materials such as superalloys and titanium, which require particularly expensive raw materials and are difficult to fabricate and machine. One important method of filling the aforementioned need is application of the relatively new powder metallurgy (P/M) technology to high-performance materials. In recent years, processes have been developed whereby superalloy and titanium aerospace hardware can be produced by the P/M process. Advantages of the P/M process include:

- The ability to produce near-net shapes, which reduces the amounts of input material and machining required to produce a finished part
- Improved property uniformity and alloy-development flexibility due to the elimination of macrosegregation
- Reduced energy requirements and shorter delivery times due to the fact that the P/M process requires fewer processing steps than conventional ingot technology.

The near-net P/M superalloy engine shaft shown in Fig. 1 is a good example of the material and fabrication savings

Fig. 1. Large P/M superalloy engine shaft made by the ceramic mold process (Ref 1)

attainable with P/M technology.[1] Figure 2 compares, schematically, the processes required for production of a shaft by cast-plus-wrought and P/M procedures.[1] It is noteworthy that in this instance the P/M process utilized markedly fewer processing steps and reduced the material input weight from 205 to 49 kg (450 to 108 lb).

The current status of P/M aerospace parts is that P/M superalloys are being used in several advanced turbine engines. These parts are being produced in shapes of varied complexity. Programs for further development of P/M superalloys for engine applications are in progress. P/M titanium is newer and is not yet in full-scale production. However, programs are in progress to develop mechanical-property data and shapemaking techniques for producing large and small complex parts for turbine-engine and airframe applications.

P/M Process

The process for producing P/M aerospace hardware generally involves production of spherical prealloyed powder, screening to remove oversize particles, blending of the

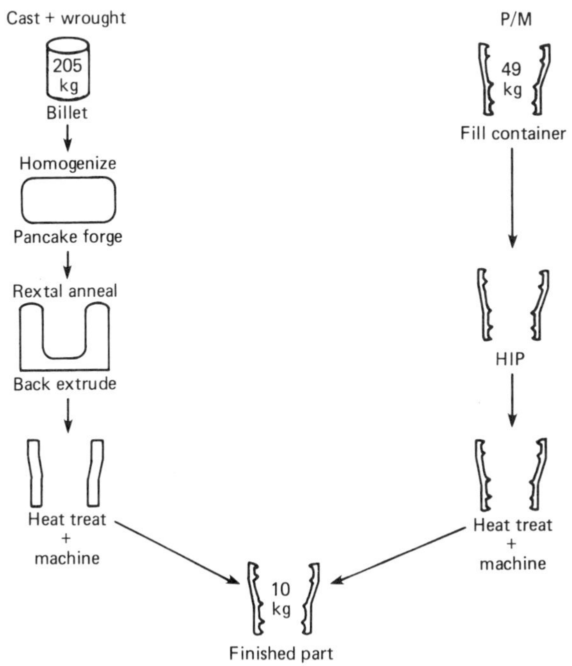

Fig. 2. Schematic comparison of processes required for manufacture of a turbine engine shaft by cast-plus-wrought and P/M procedures (Ref 1)

powders to homogenize the powder-size distribution, loading of the powder into containers, vacuum outgassing and sealing of the containers, and consolidation of the powder to full density. A production P/M superalloy process involving inert-gas atomization and hot isostatic pressing is shown schematically in Fig. 3.

The principal commercial powdermaking processes used for aerospace materials are inert-gas atomization, vacuum atomization and the plasma rotating electrode process (PREP). In the gas-atomization process used for superalloys, metal is vacuum induction melted and subsequently atomized using an inert gas—usually argon. An example of an inert-gas atomizer is shown in Fig. 4. In the vacuum-atomization powder process for superalloys, molten metal

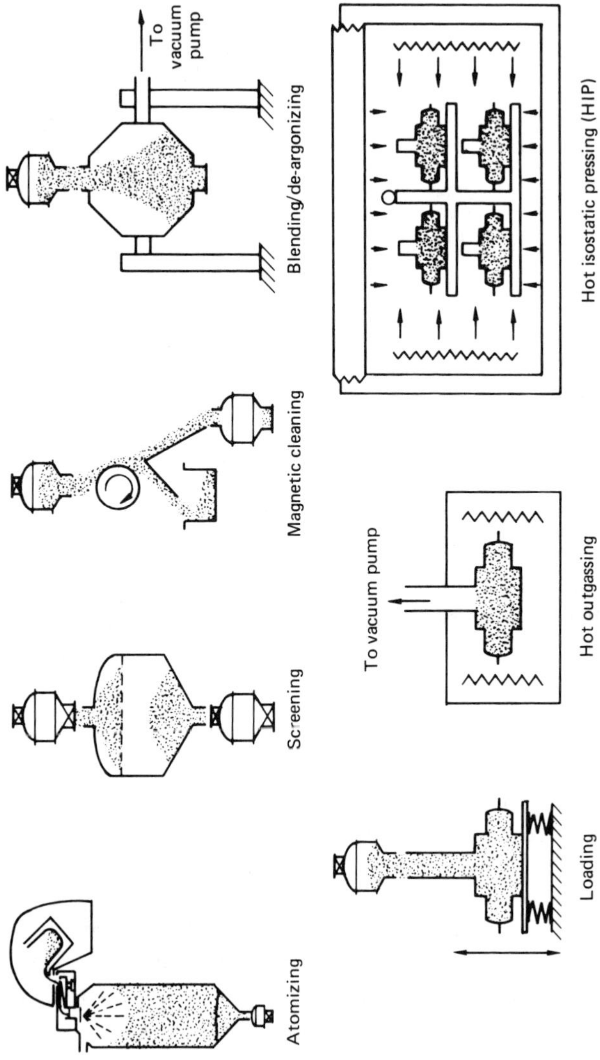

Fig. 3. Schematic illustration of a production P/M superalloy process

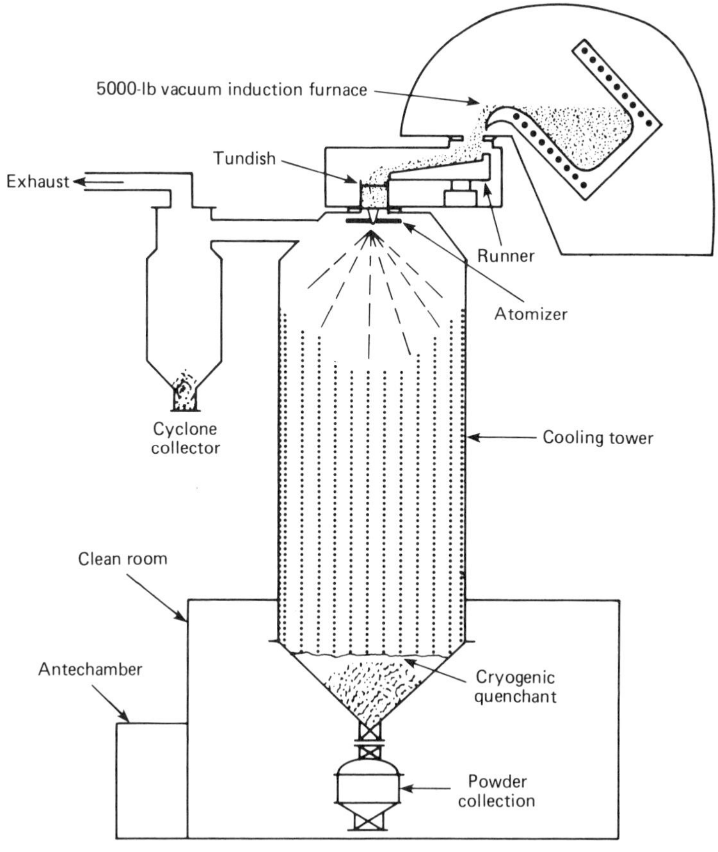

Fig. 4. Schematic illustration of a production inert-gas atomizer

is saturated with hydrogen. The metal is then streamed into a vacuum. As the gas expands, it comes out of solution and atomizes the metal. A schematic of the vacuum-atomization process is shown in Fig. 5. The plasma rotating electrode process (PREP) can be used for both superalloy powder and titanium powder.[2,3] In this process, a plasma arc provides localized melting at the end of a rapidly rotating bar. The rotating motion spins the powder off from the molten pool. This process is shown schematically

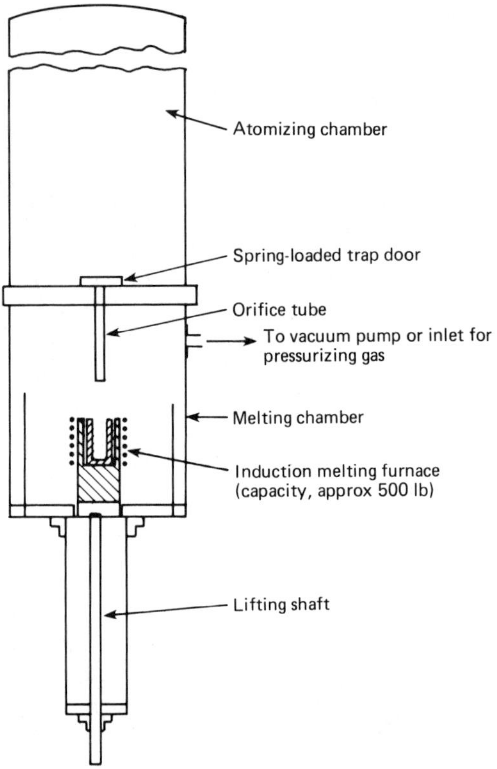

Fig. 5. Schematic illustration of a vacuum atomization plant (Ref 44)

in Fig. 6. Much of the work on P/M titanium has been done with REP (rotating electrode process) powder wherein the heat source was a tungsten arc instead of the plasma arc used in the PREP process. The original REP powder was often contaminated by tungsten particles from the electrode. This problem has been eliminated with the advent of PREP. PREP powder is also made in dedicated clean-room facilities. This has resulted in a major improvement in powder cleanness compared with REP powder.

Another atomization process that is still in the pilot stage is centrifugal atomization, which employs a rapidly spin-

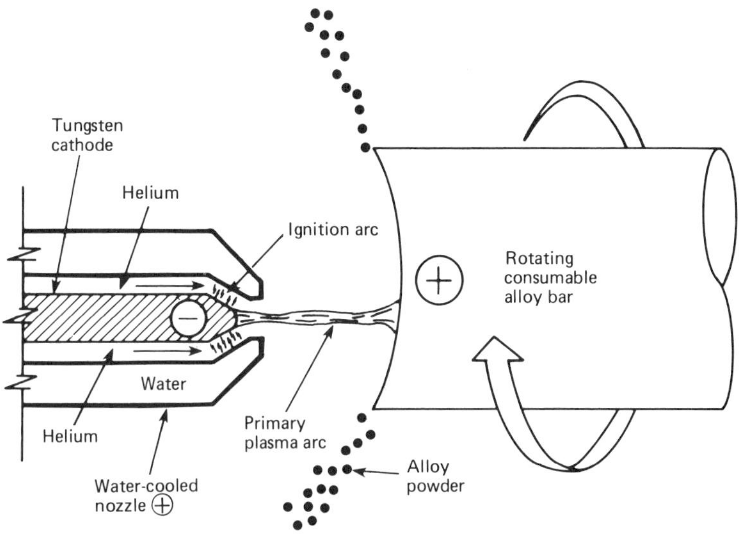

Fig. 6. Schematic illustration of the plasma rotating electrode process (PREP) (Ref 2)

ning crucible, disk or plate. In this process, molten alloy streams are poured onto a spinning disk, crucible or plate. Droplets are ejected from the spinning base. The rapid solidification process (RSP) of Pratt & Whitney Aircraft is one form of centrifugal atomization.[4,5]

Following powder production, the powder is screened to remove oversize particles. For superalloys, powder sizes range from −60 mesh (−250 μm) to −325 mesh (−44 μm) depending on the application. With superalloys, screening is utilized to limit the maximum nonmetallic inclusion size in the final product. This is a unique feature of the P/M process. Titanium alloy powders are normally screened to −35 mesh (−500 μm). After being screened, powders are subsequently blended to obtain uniform size distribution and are then ready for loading into containers for subsequent consolidation.

Metal containers are currently the main type of container used for production of P/M superalloy hardware. These may be made from simple pipe, with the ends capped (Fig. 7); from concentric pipe for producing hollow bars (Fig. 8); or from formed sheet metal for either relatively simple sonic shapes (Fig. 9) or complicated shapes (Fig. 10). The containers are loaded under a vacuum or in air and are subsequently evacuated and sealed.

Crucible has developed a unique process for producing very complicated shapes using a ceramic mold.[6,7,8] The mold is produced by the lost-wax process, which uses specially developed materials. The mold is filled with

Fig. 7. HIP'ed P/M billet made by use of a metal container

Fig. 8. Hollow bar produced by the HIP'ed powder process

Fig. 9. Sectioned metal container for a small HIP'ed turbine engine disk currently in production

Fig. 10. Sectioned metal container for a HIP'ed powder multicomponent compressor spool

powder and subsequently placed inside a cylindrical metal container surrounded by a granular ceramic (Fig. 11). The metal container is then sealed. During consolidation, the granular ceramic transmits pressure from the outer container to the ceramic container. This process is capable of producing a variety of very complex shapes such as those shown in Fig. 12.

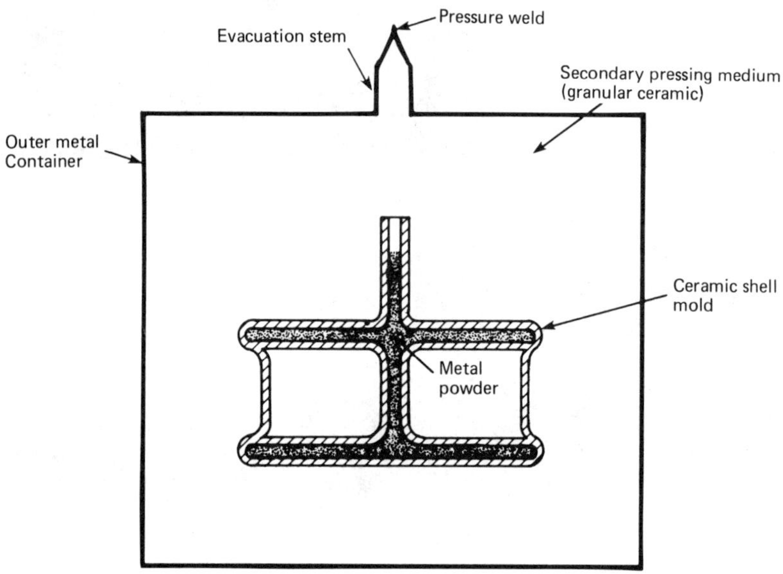

Fig. 11. Schematic illustration of a container assembly for the ceramic mold process

Fig. 12. Assorted parts made by the ceramic mold process

Other containerization methods that are in the experimental stage include (*a*) the Kelsey-Hayes fluid die process[9,10] wherein a shaped cavity is machined, cast or forged into mating metal blocks that are filled with powder and outgassed before sealing, (*b*) the Kelsey-Hayes glass bag process[11] and (*c*) the Universal-Cyclops CAP® process[12,13,14] wherein a glass container is filled with powder, outgassed, and sealed prior to compaction.

P/M superalloy and titanium alloy powders are consolidated to full density using combinations of temperature and pressure. The two principal techniques used to consolidate powder for aerospace applications are hot isostatic pressing (HIP) and extrusion followed by isothermal forging. For HIP, powder-filled containers are placed in an autoclave which is subsequently heated and pressurized. Superalloys are normally HIP'ed at temperatures ranging from 2000 to 2200°F under a pressure of 15,000 psi. Titanium alloys are normally HIP'ed at temperatures ranging from 1600 to 1800°F. Products consolidated by HIP are fully dense. The extrusion plus isothermal forging process was developed by Pratt & Whitney Aircraft and is designated GATORIZING™ (Ref 15 and 16). The extrusion process develops a fine, uniform grain structure which is superplastic at subsequent forging temperatures.

In the CAP® (consolidation by atmospheric pressure) process (currently in development), consolidation occurs by atmospheric pressure at high temperature.[12,13,14] Material produced by the CAP® process is usually not fully dense and requires subsequent hot isostatic pressing and/or hot working to attain full density.

Once consolidated to full density, such as by HIP, P/M materials undergo conventional finishing processes such as heat treatment, nondestructive inspection and machining. In many instances, however, modified heat treatments have been developed for P/M products. Also, machining requirements are often markedly reduced and inspectability of the P/M product is often enhanced. For example, the uniform structures of P/M products are ideally suited to ultrasonic inspection.

In some instances, HIP'ed material may be subsequently forged with the intent of work hardening and removing or reducing material defects. The benefits of such operations are the subject of some controversy. For example, König, Track and Betz of MTU have expressed concern that "... there is a danger that forging will introduce defects into material or that existing defects will be increased in size resulting in a reduction of strength".[45]

P/M Superalloys

The principal nickel-base superalloys being used in aerospace applications are René 95, low-carbon Astroloy, IN-100 and MERL 76 (nominal compositions are given in Table 1). With the exception of MERL 76, these alloys are slight compositional variations of cast and wrought materials. MERL 76 was developed specifically for powder application.

Powder Characteristics

Most superalloy powder produced currently is spherical and has a tap density of about 65%. Tap density is important for reliable production of P/M shapes. Figure 13 shows powder produced by argon-gas atomization. Large particles, in addition to being spherical, often have smaller satellite particles attached. Also, particles may have partial coatings of splat which result from collision of solid

Table 1. Compositions of P/M superalloys

	Composition, wt%			
Alloy	IN-100	René 95	MERL 76	Low-carbon Astroloy
Carbon	0.07	0.06	0.02	0.03
Chromium	12.5	13.0	12.4	15.0
Cobalt	18.5	8.0	18.5	17.0
Molybdenum	3.2	3.5	3.2	5.0
Tungsten	...	3.5	...	...
Hafnium	...	...	0.4	...
Niobium	...	3.5	1.4	...
Aluminum	5.0	3.5	5.0	4.0
Titanium	4.3	2.5	4.3	3.5
Vanadium	0.75	...	...	...
Boron	0.02	0.01	0.02	0.025
Zirconium	0.04	0.05	0.06	0.04
Nickel	Rem	Rem	Rem	Rem

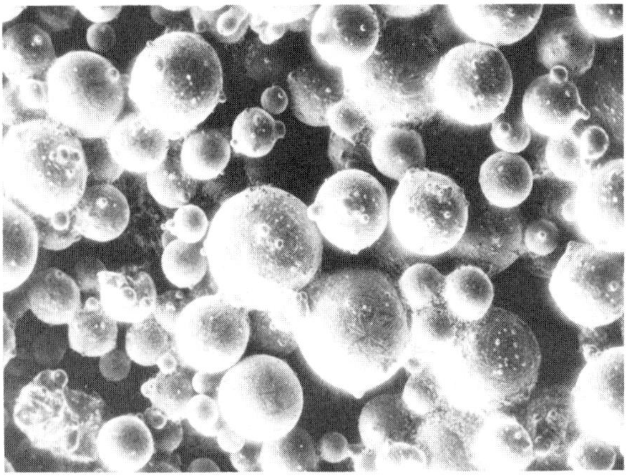

Fig. 13. Argon-atomized superalloy powder

particles with liquid particles. As shown in Fig. 14, gas-atomized powder exhibits a variety of structures.(17) Van Stone *et al* studied the microstructure of gas-atomized René 95 as a function of powder size and atomization gas.(18) Two types of structure were observed: dendritic, and nondendritic or cellular. Examples are shown in Fig. 15. Argon- and nitrogen-atomized powder had mainly dendritic microstructures, whereas helium-atomized René 95 had a primarily nondendritic structure.

Measurements of dendrite-arm spacing (DAS) are made to evaluate the relative solidification rates of powders. It is generally accepted that DAS decreases as cooling rate increases according to the following relationship:(19)

$$d = a \cdot \epsilon_{\mathrm{avg}}^{-n}$$

where d is dendrite-arm spacing, ϵ_{avg} is cooling rate during solidification and a and n are constants. It is also generally accepted that the major method for increasing cooling rate during solidification, beyond about 10^2 K/s, is to increase the ratio of surface area to volume (i.e., decrease the particle size of spherical powders).(4) The effects of powder size and atomization gas on DAS are shown in

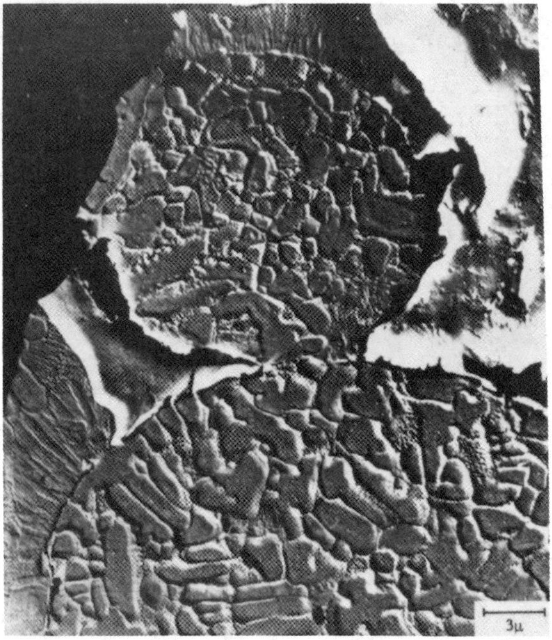

Fig. 14. Microstructure of gas-atomized IN-100 powder (Ref 17)

Fig. 16 and 17. The data show that DAS decreases, and thus cooling rate increases, with decreasing powder size. The effects of atomization gas (argon, helium or nitrogen) are relatively minor. Nitrogen- and helium-atomized powder had similar DAS, and both had finer DAS than argon-atomized powder. The cooling rates of superalloy powders can be estimated from DAS using the relationship given above and values for constants for superalloys reported by Mehrabian.[19] For the range of DAS shown in Fig. 16 and 17, the calculated cooling rates range from 10^2 to $>10^5$ K/s. This is in contrast to cooling rates for ingot metallurgy on the order of 10^{-3} to 10^{-1} K/s. Rapid solidification processing (RSP) has been defined as including cooling rates greater than 10^2 K/s.[20] Gas-atomized powders, therefore, are classified as RSP materials.

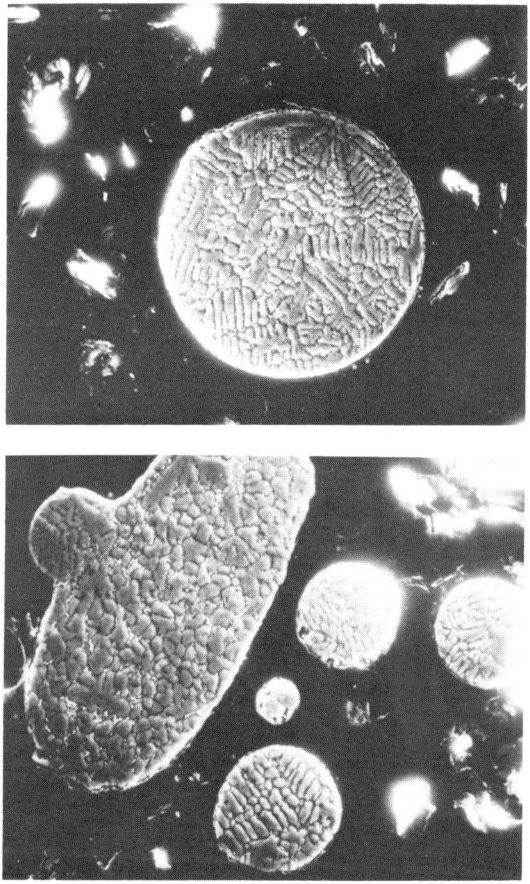

(Top) Dendritic structure. Magnification, 2000×. (Bottom) Nondendritic or cellular structure (elongated particle). Magnification, 2000×.

Fig. 15. Microstructures of gas-atomized superalloy powder (Ref 21)

Upon consolidation, certain superalloy powders develop carbide networks at prior particle boundaries. It is believed that this characteristic is the result of the carbon content and the amount and type of carbide-forming elements

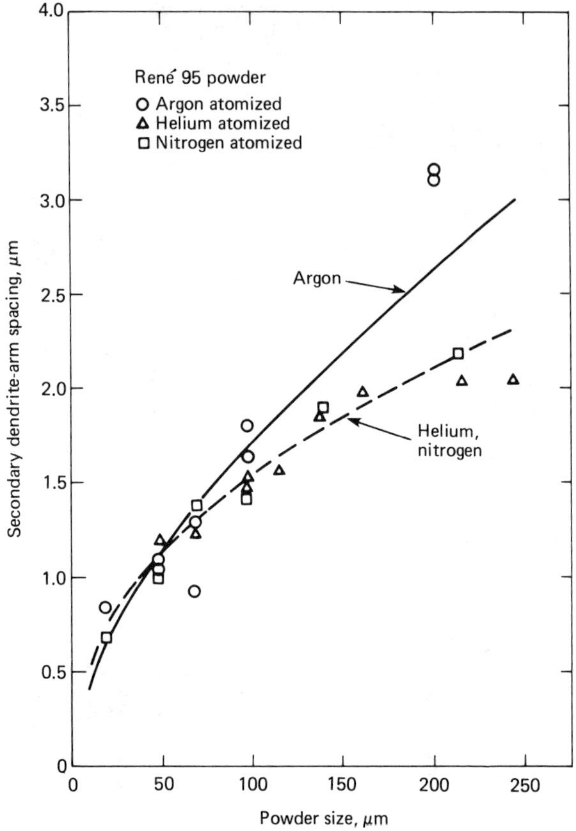

Fig. 16. Variation of dendrite-arm spacing with powder size for gas-atomized René 95 (Ref 18)

present in the alloy. For example, the casting composition of alloy IN-100 (0.18 C, 10.0 Cr, 15.0 Co, 5.0 Ti, 5.5 Al, 1.0 V, 0.05 Zr, 0.015 B, rem Ni) exhibits heavy outlining of carbide particles when produced as a powder and HIP'ed, but P/M alloy IN-792 (0.21 C, 12.7 Cr, 9.0 Co, 2.0 Mo, 3.9 Ta, 3.9 W, 4.2 Ti, 3.2 Al, 0.10 Zr, 0.02 B, rem Ni), also a casting alloy, does not develop substantial amounts of carbide at prior particle boundaries (Fig. 18).

To develop a better understanding of the carbide reactions that occur during heating for hot compaction, the

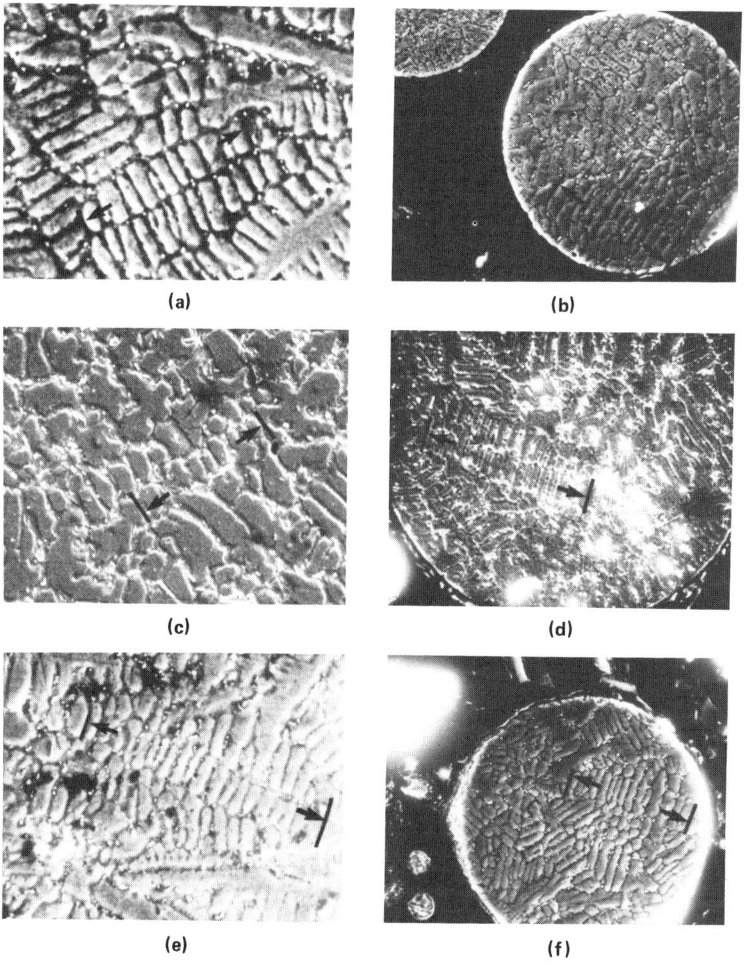

(a) Argon-atomized powder; −140 +170 mesh (−100 +88 μm); DAS = 1.73 μm. (b) Argon-atomized powder; −400 mesh (−37 μm); DAS = 1.07 μm. (c) Nitrogen-atomized powder; −140 +170 mesh (−100 +88 μm); DAS = 1.87 μm. (d) Nitrogen-atomized powder; −400 mesh (−37 μm); DAS = 0.89 μm. (e) Helium-atomized powder; −140 +170 mesh (−100 +88 μm); DAS = 1.35 μm. (f) Helium-atomized powder; −400 mesh (−37 μm); DAS = 0.85 μm.

Fig. 17. Typical dendrite-arm spacing (DAS) values and powder structures resulting from various atomization processes for two powder sizes (Ref 21)

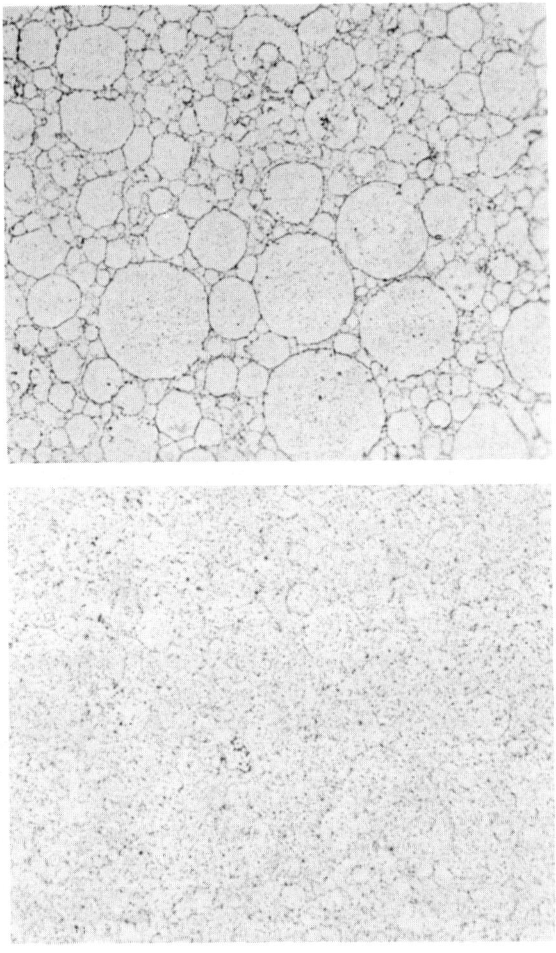

(Top) P/M alloy IN-100 (casting composition). (Bottom) P/M alloy IN-792. Both specimens were etched with Murakami's reagent to stain carbides.

Fig. 18. Carbide distributions in powder particles of two casting alloys (Ref 21)

carbide reactions in alloys IN-100 and IN-792 were investigated by heating the powders to various temperatures and then metallographically examining the resulting carbide distributions.[21] Photomicrographs showing the carbide distributions in IN-100 and IN-792 powders are presented in Fig. 19 and 20, respectively. Note that the powders were etched using Murakami's reagent, which stains carbides but does not attack the matrix or other secondary phases. It can be noted that both alloys contain relatively uniform distributions of carbide in the as-atomized condition. When heated to temperatures ranging from 1950 to 2350°F, the carbide distribution in IN-100 (Fig. 19) gradually changes from a uniform distribution to a concentration at the particle surface resulting from the carbide going into solution and precipitating out at the surface. The roughing of the particle surface is due to stained carbides at the surface. In contrast, IN-792 (Fig. 20) develops little or no carbide at the particle surface. The carbide within the particle gradually coarsens with increasing temperature. At 2350°F, some melting occurs and carbides appear to be associated with the melted areas.

The difference in carbide behavior between IN-100 and IN-792 is believed to be related to differences in composition, which affect the type and amount of carbide present. The carbide formed in IN-100 during solidification is relatively unstable (e.g., $M_{23}C_6$) and goes into solution at elevated temperatures. As carbon becomes available, a more stable carbide (e.g., MC) forms at the particle surface, which is an energetically favorable nucleation site. The carbide formed in IN-792 during solidification (e.g., MC) is stable up to the incipient melting temperature, and sufficient carbon is not available for precipitation at the particle surface. It is believed that the difference in behavior of these two alloys is attributable to the presence of a substantial amount of tantalum in IN-792. Tantalum is a very potent MC former. As a result, stable MC forms during solidification instead of a relatively unstable carbide such as $M_{23}C_6$. Niobium and hafnium are believed to have similar effects on carbide precipitation. Thus, two means are available for limiting the formation of particle-surface carbide in superalloy powders. One is to simply lower the carbon content; this is effective with IN-100 and Astroloy. The other is to modify the composition to include strong MC carbide-forming elements such as tantalum, niobium and hafnium.

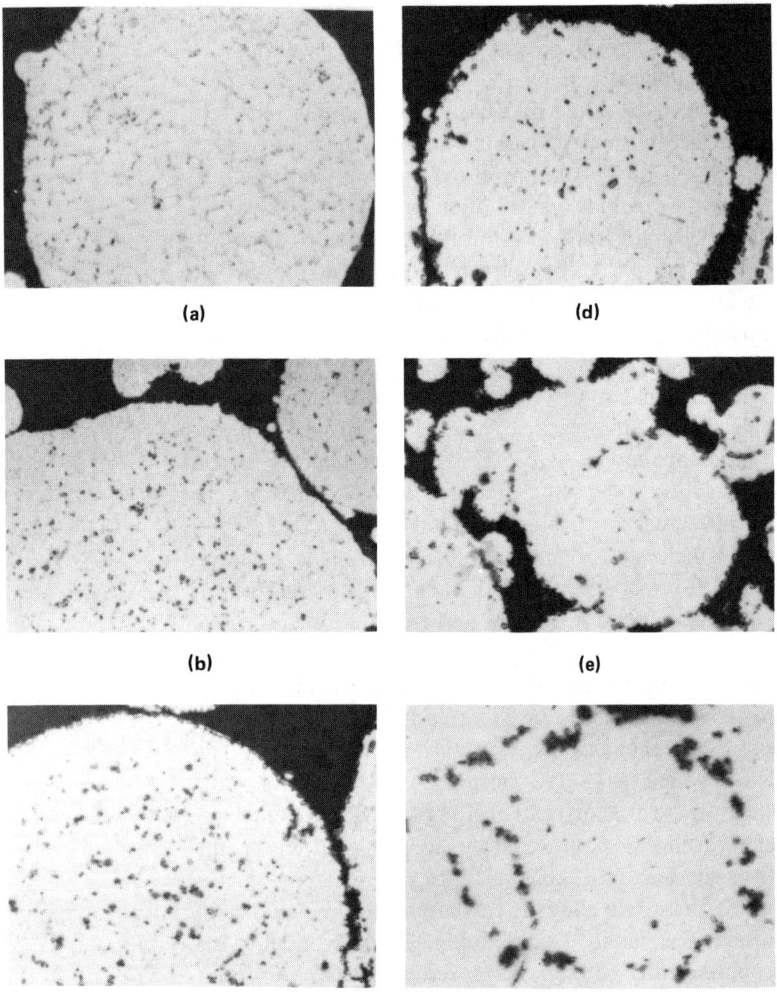

(a) (d)

(b) (e)

(c) (f)

(a) As-atomized. (b) Exposed for 4 h at 1950°F. (c) Exposed for 4 h at 2050°F. (d) Exposed for 4 h at 2150°F. (e) Exposed for 4 h at 2250°F. (f) Exposed for 4 h at 2350°F. All specimens were etched with Murakami's reagent to stain carbides. Magnification, 1000×.

Fig. 19. Effect of thermal exposure on carbides in alloy IN-100 powder (Ref 21)

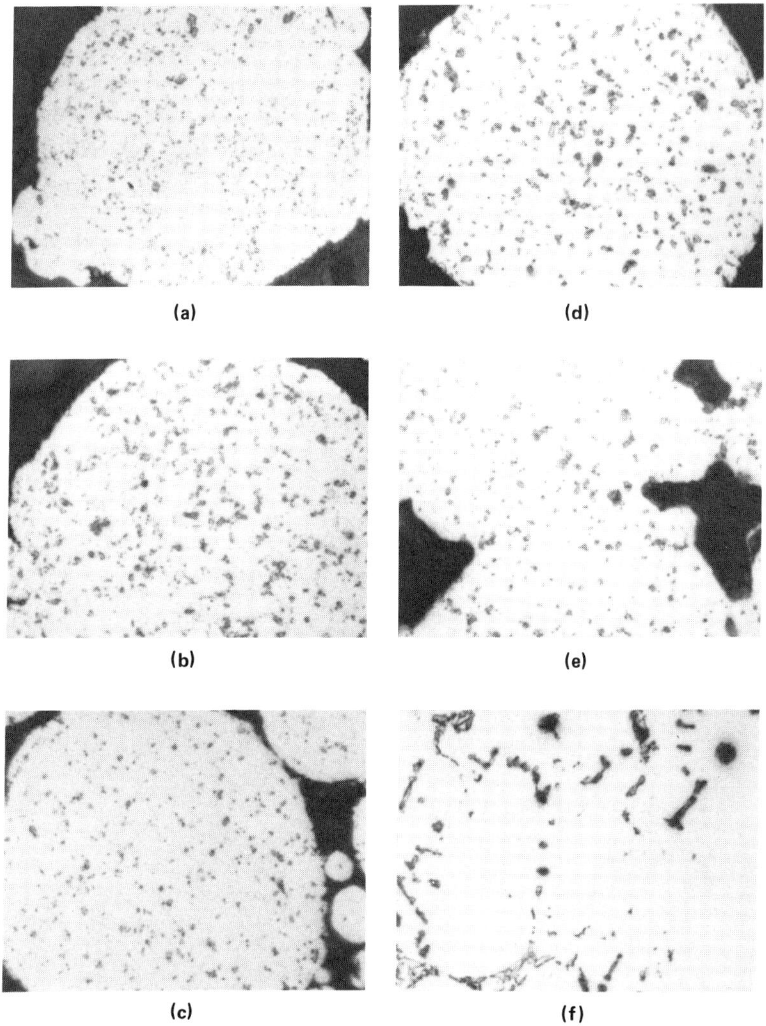

(a) As-atomized. (b) Exposed for 4 h at 1950°F. (c) Exposed for 4 h at 2050°F. (d) Exposed for 4 h at 2150°F. (e) Exposed for 4 h at 2250°F. (f) Exposed for 4 h at 2350°F. All specimens were etched with Murakami's reagent to stain carbides. Magnification, 1000×.

Fig. 20. Effect of thermal exposure on carbides in alloy IN-792 powder (Ref 21)

Mechanical Properties

René 95 was originally developed by General Electric as a high-strength cast-plus-wrought alloy for turbine-engine applications. Early work showed the alloy to be well suited for manufacture by the powder process.[22,23] The composition of the P/M version of this alloy is essentially the same as the cast-plus-wrought material except that the carbon content is slightly lower in the P/M material. P/M René 95 is currently used in turbine engines in the HIP'ed and HIP'ed-plus-forged conditions. Table 2 compares the properties of P/M René 95 with those of conventional cast-plus-wrought material.[16] Generally, the properties of

Table 2. Mechanical properties of René 95 turbine wheels made by various processes(a)

Property	Cast and forged	Powder preform forged	HIP'ed and heat treated	Specification
Room-Temperature Tensile Properties				
Tensile strength, ksi	215	240	240	225
Yield strength, ksi	187	182	180	163
Elongation, %	7	15	20	10
Reduction in area, %	8	16	21	12
Tensile Properties at 1200°F				
Tensile strength, ksi	208	220	218	205
Yield strength, ksi	177	171	164	153
Elongation, %	7	14	13	8
Reduction in area, %	7	17	14	10
Stress-Rupture Properties at 1200°F and 150 ksi				
Life, h	70	93	54	50
Elongation, %	4	9	5	3
Creep Properties at 1100°F and 150 ksi				
Time, h	100	100	100	50
Strain, %	0.29	0.22	0.14	0.3 max

(a) Property values are averages of data given in Ref 16.

HIP'ed and HIP'ed-plus-forged materials are similar. Compared to cast plus wrought, the P/M product exhibits slightly lower yield strength but markedly better tensile and stress-rupture ductility.

Statistical mechanical-property data for 150 as-HIP'ed P/M René 95 turbine disks (Table 3) show that the properties of these disks substantially exceed specification requirements and exhibit excellent uniformity.(7) Figure 21 shows a near-net as-HIP'ed CF6-50 HPT aft shaft made by the ceramic mold process. A schematic representation of the near-net and finish-machined shapes for this part is shown in Fig. 22. Mechanical-property data for the René 95 as-HIP'ed shaft (Table 4) show that the data met goal properties.(7)

Astroloy was originally developed as a cast or cast-plus-wrought material. Initial evaluations of this alloy produced as powder showed it to be prone to carbide decoration at prior particle boundaries. Subsequent evaluations showed

Table 3. Cut-up mechanical properties of as-HIP'ed P/M René 95 turbine disks(a) (Ref 7)

Property	Specification	Mean	σ
Room-Temperature Tensile Properties			
Tensile strength, ksi . . .	225.0	239.4	2.8
Yield strength, ksi	163.0	176.9	3.5
Elongation, %	10	17.2	1.9
Reduction in area, %. . .	12	20.5	1.8
Tensile Properties at 1200°F			
Tensile strength, ksi . . .	205.0	220.1	5.4
Yield strength, ksi	153.0	165.2	4.9
Elongation, %	8	13.8	4.1
Reduction in area, %. . .	10	16.5	3.3
Stress-Rupture Properties at 1200°F and 150 ksi			
Life, h	50	88.8	26.3
Elongation, %	3.0	6.5	2.1
Creep Properties at 1100°F and 150 ksi			
Time, h.	50	54.6	6.7
Elongation, %	0.3 max	0.10	0.08

(a) First 75 engine sets (150 disks).

Fig. 21. As-HIP'ed P/M René 95 CF6-50 HPT engine aft shaft (see Fig. 22)

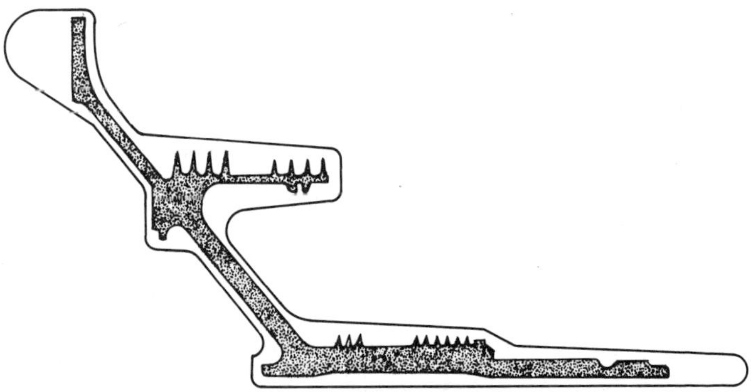

Fig. 22. Schematic representation of the engine aft shaft shown in Fig. 21, illustrating both near-net and finish-machined shapes (Ref 17)

Table 4. Tensile properties of test rings from as-HIP'ed René 95 CF6-50 HPT aft shaft(a) (Ref 17)

Test ring	Specimen location	Testing temperature, °F	Yield strength, ksi	Tensile strength, ksi	Elongation, %	Reduction in area, %
Large-diam	0°	RT	175	222	11	11
	180°	RT	176	237	16	15
	180°	RT	175	227	13	13
Small-diam	0°	RT	173	239	16	14
	180°	RT	175	235	15	14
Large-diam	0°	1200	159	202	8	12
	180°	1200	160	200	11	12
Small-diam	0°	1200	159	203	7	9
	180°	1200	162	209	14	13
	180°	1200	160	213	14	16
Goal	...	RT	150	185	10	12
	...	1200	125	145	8	10

(a) Heat treatment: 2000°F/1 h/RAC + 1400°F/16 h/AC.

that the properties of P/M Astroloy with lower-than-normal carbon levels were attractive.[24,25] Table 5 presents data for HIP'ed Astroloy at two carbon levels and two HIP temperatures. The HIP temperatures used were 2050°F, which was below the γ' solvus, and 2100°F, which was above the γ' solvus. The data show that lower carbon content resulted in marked improvements in both tensile and stress-rupture ductility and only relatively small reductions in tensile and yield strengths. HIP temperature also has a marked effect on properties. HIP'ing below the γ' solvus maximizes tensile and yield strengths but results in relatively low stress-rupture life and ductility. HIP'ing above the γ' solvus improves stress-rupture life and ductility but results in small decreases in tensile and yield strengths. In practice, Astroloy powder is melted to low-carbon specifications and HIP'ed above the γ' solvus to develop an optimum combination of properties.

IN-100 was originally developed as a casting alloy, but it is now widely used at Pratt & Whitney Aircraft as a P/M material. The composition was modified for P/M application.[15] P/M IN-100 hardware is made primarily by GATORIZING™ of extruded billet. Properties of this

Table 5. Mechanical properties of as-HIP'ed P/M Astroloy for two carbon levels and two HIP temperatures(a) (Ref 21)

Property	Low carbon (0.02-0.04%) 2050°F HIP	Low carbon (0.02-0.04%) 2100°F HIP	Standard carbon (0.03-0.09%) 2050°F HIP	Standard carbon (0.03-0.09%) 2100°F HIP
Room-Temperature Tensile Properties				
Tensile strength, ksi	210	200	200	190
Yield strength, ksi	145	135	155	150
Elongation, %	30	30	15	10
Reduction in area, %	30	30	15	15
Tensile Properties at 1000°F				
Tensile strength, ksi	200	185	...	...
Yield strength, ksi	135	130	...	...
Elongation, %	25	20	...	...
Reduction in area, %	25	20	...	...
Tensile Properties at 1200°F				
Tensile strength, ksi	185	180	...	...
Yield strength, ksi	130	120	...	...
Elongation, %	25	35	...	...
Reduction in area, %	25	35	...	...
Tensile Properties at 1400°F				
Tensile strength, ksi	145	145	150	150
Yield strength, ksi	130	115	135	135
Elongation, %	15	25	5	5
Reduction in area, %	15	25	10	10
Stress-Rupture Properties at 1350°F and 80 ksi				
Smooth life, h	95	200	...	...
Elongation, %	5	12	...	...
Reduction in area, %	10	15	...	...
Notched life, h(b)	60	280	...	...
Stress-Rupture Properties at 1400°F and 85 ksi				
Smooth life, h	5	30	8	13
Elongation, %	8	15	3	4
Reduction in area, %	8	15	3	3
Notched life, h(b)	10	55	...	...
Creep Property at 1300°F and 74 ksi				
Time to 0.1% plastic deformation, h	100	150	...	...

(a) Specimens taken from material 3 in. in diameter by 3 in. long, heat treated as follows: 2050°F/4 h/AC + 1600°F/8 h/AC + 1800°F/4 h/AC + 1200°F/24 h/AC + 1400°F/8 h/AC. (b) $K_t = 3.7$.

Table 6. Mechanical properties of P/M IN-100 disk made by two processes(a)(b)

Property	Extruded plus GATORIZED™	HIP plus GATORIZED™
Room-Temperature Tensile Properties		
Tensile strength, ksi	231	231
Yield strength, ksi	160	159
Elongation, %	26	26
Reduction in area, %	26	28
Tensile Properties at 1300°F		
Tensile strength, ksi	184	184
Yield strength, ksi	155	152
Elongation, %	18	19
Reduction in area, %	20	23
Stress-Rupture Properties at 1350°F and 95 ksi		
Life, h	35	33
Elongation, %	13	14
Creep Properties at 1300°F and 80 ksi		
Time to 0.1%, h	102	110
Time to 0.2%, h	134	154

(a) Property values are averages of data given in Ref 16. (b) Heat treatment: 2050°F/2 h/OQ + 1600°F/40 min/AC + 1800°F/45 min/AC + 1200°F/24 h/AC + 1400°F/4 h/AC.

material are given in Table 6.[16] Also included are data for HIP billet that was subsequently GATORIZED™. The properties are generally similar to those developed in extruded billet.

MERL 76 is an alloy developed by Pratt & Whitney Aircraft for fabrication by hot isostatic pressing of prealloyed powder into engine components.[26] It was designed to have mechanical properties equal to those of isothermally forged IN-100. The composition of MERL 76 is based on the IN-100 alloy but with lower carbon and with additions of niobium and hafnium. Reportedly, these compositional modifications were required for ductility and carbide control.

Mechanical-property data for as-HIP MERL 76 have been reported by Evans and Eng.[26] The MERL 76 was HIP'ed

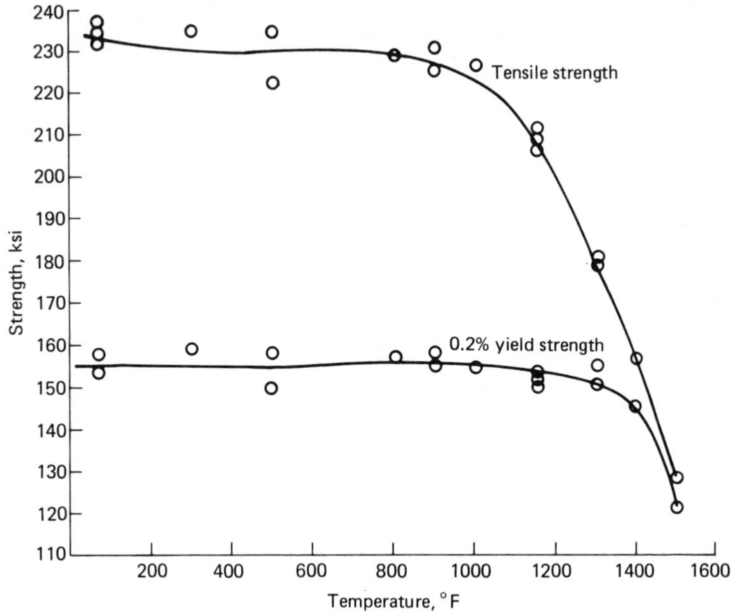

Fig. 23. Tensile strength vs temperature for as-HIP'ed P/M alloy MERL 76 (Ref 26)

at 2160°F and subsequently heat treated as follows: 2125°F/2 h/OQ + 1600°F/40 min/AC + 1800°F/45 min/AC + 1200°F/24 h/AC + 1400°F/16 h/AC. The resulting tensile and yield strengths are plotted as functions of temperature in Fig. 23. It is noteworthy that the yield strength of the alloy is relatively constant up to about 1250°F. This feature is probably the result of the high γ' content (65 vol %) of the alloy. Tensile properties for different section thicknesses are presented in Table 7. Generally, strength decreases slightly and ductility increases slightly with increasing section size. This effect is the result of differences in cooling rate between the various section sizes during heat treatment. Stress-rupture properties and 1000°F axial low-cycle fatigue properties of MERL 76 are shown in Fig. 24 and 25, respectively. Reportedly, MERL 76 met all minimum requirements for its intended engine-disk application.

Table 7. Effect of section size on tensile properties of MERL 76 (Ref 26)

Section thickness, in.	Temperature, °F	Yield strength(a), ksi	Tensile strength, ksi	Elongation, %	Reduction in area, %
1.5	70	156	233	20	28
	1150	154	215	21	28
	1300	152	191	16	23
3.0	70	153	228	25	23
	1150	147	205	28	33
4.5	70	148	230	20	22
	1150	147	209	23	28
	1300	146	183	21	27

(a) At 0.2% offset.

P/M Superalloy Applications

P/M superalloys are currently being used or developed for use in a variety of turbine-engine components.[2,15,27] These include turbine disks, compressor disks, spacers, cooling plates and shafts. Engine systems for which P/M

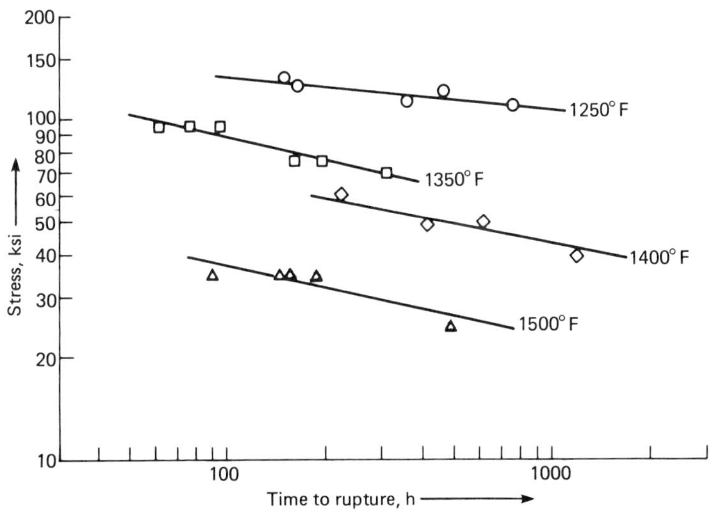

Fig. 24. Stress-rupture properties of as-HIP'ed P/M alloy MERL 76 (Ref 26)

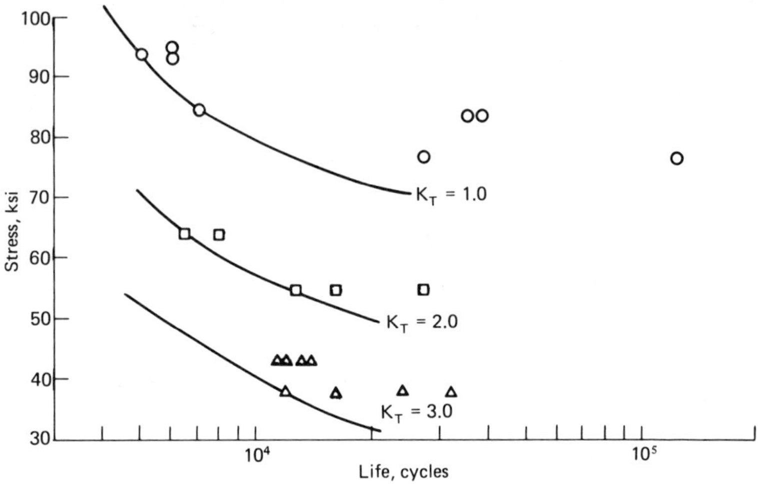

Fig. 25. Axial low-cycle fatigue life of as-HIP'ed P/M alloy MERL 76 at 1000°F (Ref 26)

superalloy components are being used or being considered for use include General Electric's T-700, F-404 and CFM-56 and Pratt & Whitney's JT8D, JT9D and F-100. These components range in size from relatively small turbine disks (Fig. 26) to large turbine components (Fig. 27).

The P/M process is also being developed to produce multicomponent parts.[8] Using conventional technology, engine spools are normally assembled from parts made from several individual forgings. With the P/M process, the spools can be made as a single piece. The result is a marked reduction in input weight and the elimination of a number of processing steps. A sectioned container for a complex spool is shown in Fig. 10. In this example, the spool is comprised of six segments: a shaft at the top, four compressor disks, and a bolt flange at the bottom. The total input weight of individual components is 490 lb. If the spool is made as a single component, the total input weight is 210 lb, and thus a material savings of 280 lb per engine is realized by using P/M multicomponent technology.

Fig. 26. Small production P/M turbine disks made by use of a metal container

Fig. 27. Large P/M turbine disk

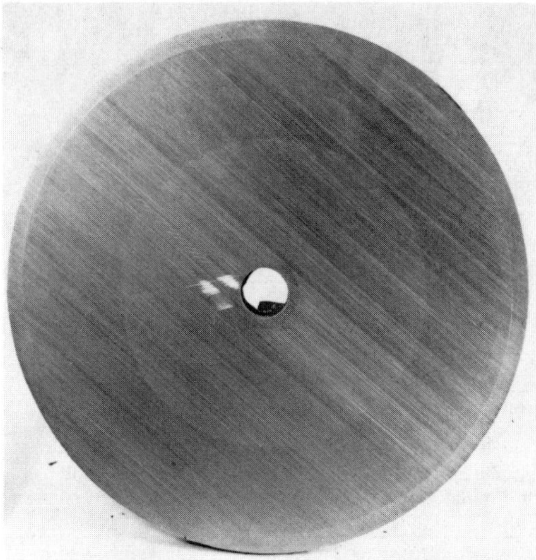

Fig. 28. P/M shaper cutter blank (mild steel core, M2 tool steel case) (Ref 8)

P/M technology is being developed to produce dual-alloy, dual-property components. Such a component may simply be a composite bar such as that shown in Fig. 28.[8] This is a shaper cutter blank made by bonding M2 tool steel to the outside surface of a mild steel core. Several programs have been conducted or are in progress to develop dual-property turbine disks. By use of these techniques, the high tensile and low-cycle fatigue property requirements of the bore can be attained with a fine-grain P/M product and the creep and stress-rupture requirements of the rim and/or blades can be attained with special processing techniques (e.g., single-crystal casting, directionally solidified casting, fiber reinforcement or mechanical alloying). The ceramic mold process is being developed to produce dual-property turbine wheels for small engines.[28] One technique is shown schematically in Fig. 29. This concept, studied by Avco-Lycoming, uses a blade ring made by tack welding of individual cast alloy (C103) blades. The ring is then sandwiched between two halves of a ceramic mold.

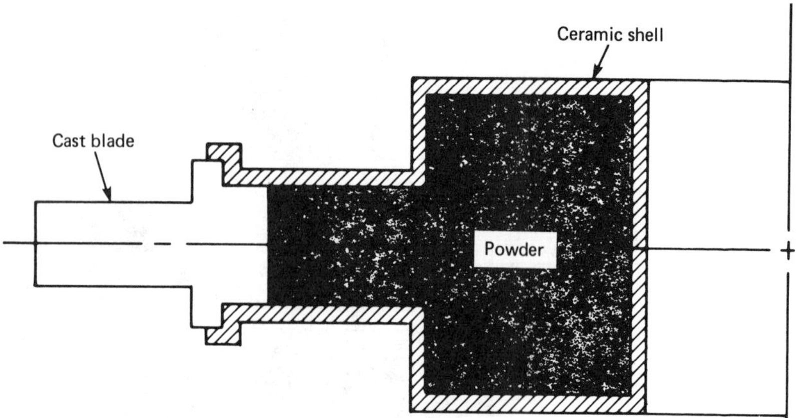

Fig. 29. Schematic illustration of a P/M dual-property turbine wheel process

The mold is then filled with powder (alloy PA101) and the assembly is HIP'ed to consolidate the powder and to bond the powder to the blades. Figure 30 shows the mold being fabricated, Fig. 31 shows the HIP'ed wheel and Fig. 32 shows the bond line between the two materials.

Another dual-property wheel technique, developed by Detroit Diesel Allison, involves the use of a preconsolidated P/M hub and a directionally solidified cast blade ring.[28] The two components are joined by shrink fitting the blade ring onto the hub, sealing the joint by brazing, and then HIP'ing to bond the components. Figure 33 shows a completed wheel. The coarse-grain directionally solidified rim (alloy MAR-M247) and a fine-grain P/M hub (alloy PA101) can be noted. Other dual-alloy techniques have been successfully explored and demonstrated using metal-container technology.[29]

P/M Titanium Alloys

Powder Characteristics

REP (Rotating Electrode Process) and PREP (Plasma Rotating Electrode Process) powders are spherical prealloyed powders (Fig. 34) with very few irregular particles

Fig. 30. Assembly of a P/M dual-property turbine wheel

Fig. 31. Dual-property turbine wheel made from cast blades and a P/M hub (Ref 28)

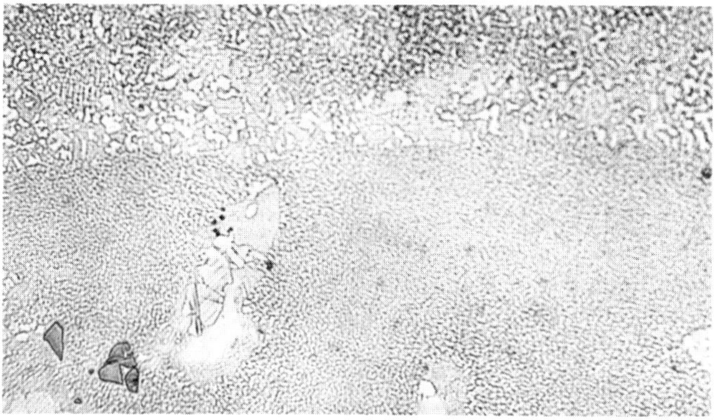

Fig. 32. Microstructure at bond line between cast alloy C103 (top) and P/M alloy PA101 (bottom) (Ref 28)

Fig. 33. Dual-property turbine wheel made by bonding a fine-grain P/M hub to a coarse-grain directionally solidified cast blade ring (Ref 28). This wheel was produced by Detroit Diesel Allison.

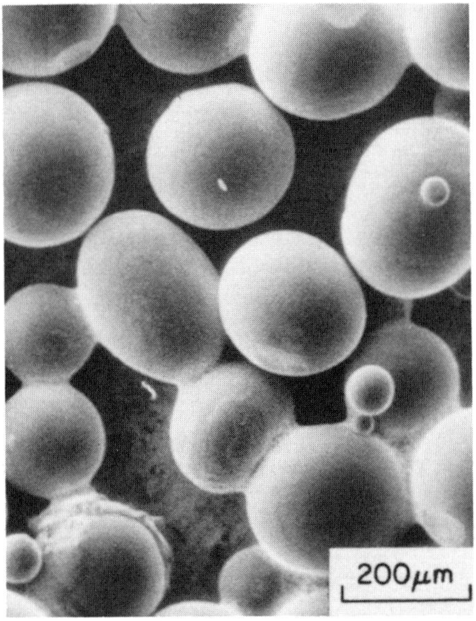

Fig. 34. Ti-6Al-4V powder made by the REP process (see also Fig. 35)

or satellite particles. The over-all size distribution is typically from −35 mesh (500 μm) to +325 mesh (44 μm). Packing density of the powder is about 65% of theoretical density and the powder has very good flow characteristics. The powder is a high-purity material with a composition virtually identical to that of the starting bar stock. Powder produced with a plasma heat source (PREP) is free of tungsten inclusions, which can be introduced when the powder is made with a tungsten electrode. No refractories are used in the process and the powder is handled under clean-room conditions, ensuring a high level of cleanness with respect to nonmetallic inclusions.

The microstructures of REP and PREP powders are typical of material rapidly cooled from the molten state. With alloy Ti-6Al-4V, the structure is martensitic with very fine needles (Fig. 35). More highly alloyed materials exhibit dendritic structures. The dendrite-arm spacing in REP

Fig. 35. Microstructure of Ti-6Al-4V powder made by the REP process

powder is reported to be on the order of 3 to 6 μm.(30) Frequency of gas pores or shrinkage cavities in REP powder is very low. The characteristics of REP and PREP powders make them excellent materials for hot consolidation of near-net shapes.

Mechanical Properties of Prealloyed Powder

The titanium alloys of major interest for prealloyed powder application arc Ti-6Al-4V Standard (Std), Ti-6Al-4V Extra Low Interstitial (ELI) and Ti-6Al-6V-2Sn. To date, the mechanical properties of P/M titanium alloys have been measured on material made from REP (tungsten arc) powder. Data for PREP (plasma arc) powder are only now being developed. Early information indicates that PREP powder is free of tungsten contamination and is cleaner than REP powder.(21)

Table 8. Room-temperature tensile and toughness properties of HIP Ti-6Al-4V (Std) powder (Ref 31)

Condition	Orientation	Yield strength, ksi	Tensile strength, ksi	Elongation, %	Reduction in area, %	W/A (a), in. · lb/in.^{2}	Predicted K_Q (b), ksi $\sqrt{in.}$
As-HIP	LT	135	142	12	40	410	63
HIP + vacuum annealed(c)	LT	132	139	13	38	565	73
	TL	130	138	12	35	578	74
	ST	132	139	12	36	541	72
HIP + beta annealed (d)	LT	131	138	9.5	23	446	65
AMS Specification 4928H	–	120	130	10	15	–	–

(a) Tested in compression mode. (b) Predicted from $K_Q^2/E - 0.57(W/A)$. (c) Annealing treatment: 1300°F/8 h/slow cool. (d) Annealing treatment: 1900°F/½ h/slow cool to 1400°F/hold 2 h/AC.

Ti-6Al-4V (Std) is the leading cast and wrought titanium alloy in terms of usage. Its popularity has carried over into P/M technology. General Dynamics conducted a preliminary evaluation of this alloy. The data[31] were developed on an alpha-plus-beta HIP'ed test block (3 by 5 by 5 in.). Tensile properties developed in the 1300°F anneal condition (Table 8) for the three principal directions (LT, TL and ST) show the test block to be isotropic. Use of a 1900°F beta anneal gave somewhat lower tensile ductility. However, the K_Q toughness values reported for both the 1300 and 1900°F anneals exceeded the typical value of 60 ksi $\sqrt{in.}$ for annealed wrought material.[32]

High-cycle fatigue (HCF) data were developed on Ti-6Al-4V (Std) by Williams International in a program to apply near-net-shape HIP technology to a compressor rotor part for the F-107 cruise missile engine. In this study,[33] two size fractions of powder were used: −35 mesh (as-received) and −80 mesh. There was no difference in HCF test results between the two sizes. Room-temperature and 700°F S/N curves are shown in Fig. 36.

Notched fatigue data were developed by the McDonnell Aircraft Company[34] on alpha-beta HIP'ed Ti-6Al-4V

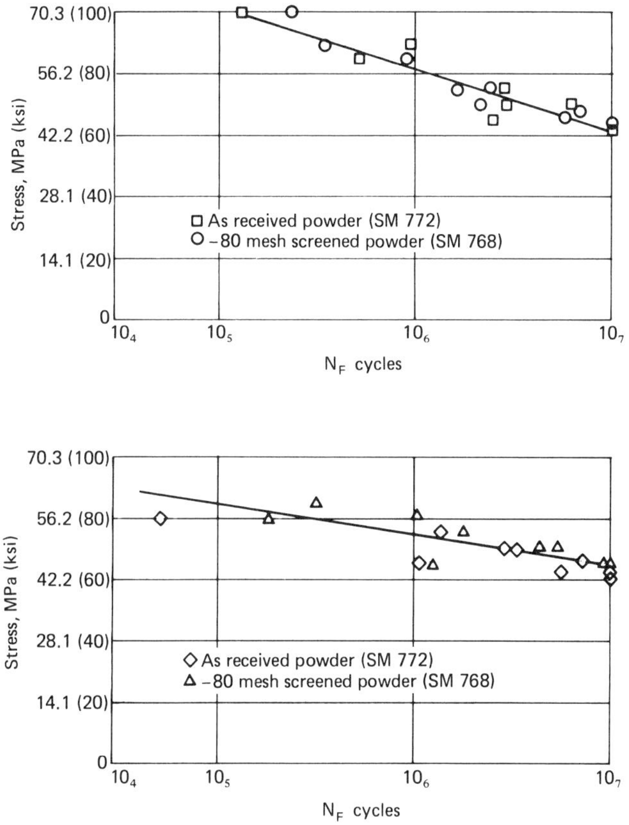

(Top) Room-temperature properties. (Bottom) Properties at 700°F.

Fig. 36. High-cycle fatigue properties of P/M Ti-6Al-4V (Std) (Ref 33)

(Std) test blocks (2 by 4 by 8 in.) in the 1300°F mill-annealed condition (Fig. 37). Here the fatigue strength of the HIP material is superior to that of alpha-beta-processed forgings, which is included for comparison.

Ti-6Al-4V (ELI) has also been evaluated as a HIP'ed P/M material. A heat treatment study using test blocks was performed by Northrop[35] to determine the optimum treatment for HIP'ed material. Tensile-test results are shown in

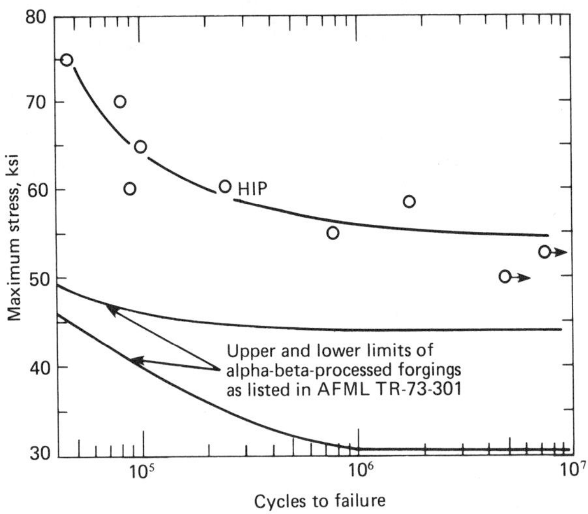

$K_t = 3$; Hz = 30; R = 0.1

Fig. 37. Notched fatigue strength of HIP'ed P/M Ti-6Al-4V (Ref 34)

Table 9. These results meet the McDonnell Aircraft Specification MMS-1225 requirement for Ti-6Al-4V (ELI) parts. In this table, the fracture-toughness values for the three annealed conditions exceed the typical value for forgings. The fatigue-crack-growth rates for material in the four conditions studied showed a relatively similar crack-growth-rate behavior except for the beta-annealed condition, which showed slightly superior fatigue-crack-growth resistance.

A fuselage brace made from HIP'ed Ti-6Al-6V-2Sn powder was used by Grumman Aerospace to establish the flight worthiness of a HIP'ed P/M airframe component.[36] The tensile and toughness properties developed by Grumman (Table 10) compare well with the average values for forgings. The fatigue endurance limit of HIP powder developed in this program is given in Fig. 38. Here the HIP data points lie within the band for annealed forgings and plate.

In the work of Dawson and Ulitchny[37] the effects of P/M processing variables and heat treatments on the micro-

Table 9. Room-temperature tensile and toughness properties of HIP Ti-6Al-4V (ELI) powder (Ref 35)

Condition	Yield strength, ksi	Tensile strength, ksi	Elongation, %	K_{IC}, ksi $\sqrt{in.}$
As-HIP	124	135	15	90(a)
HIP + beta annealed(b)	130	138	10	85
HIP + recrystallization annealed(c)	125	134	15	87(a)
HIP + mill annealed(d)	119	128	17	89(a)
Typical wrought (mill annealed)	115	125	12	84
MMS-1225 Specification (annealed)	110	120	10	70

(a) K_Q values. (b) Annealing treatment: 1300°F/12 h (vacuum degassed) + 1750°F/30 min + 1865°F/15 min/air blast cool + 1300°F/2 h/AC. (c) Annealing treatment: 1300°F/12 h (vacuum degassed) + 1759°F/2 h/AC + 1400°F/2 h/AC (to below 900°F in 45 min). (d) Annealing treatment: 1300°F/12 h (vacuum degassed) + 1300°F/2 h/FC to 1000°F (300°F/h max)/AC.

Table 10. Room-temperature tensile and toughness properties of HIP Ti-6Al-6V-2Sn powder(a) (Ref 36)

Source	Yield strength(b), ksi	Tensile strength, ksi	Elongation, %	Reduction in area, %	K_{IC}, ksi $\sqrt{in.}$
HIP	141	148	18	43	60
Forging (avg)	143	151	16	38	35-60
Grumman Specification GM3119 (min)	130	140	8	20	...

(a) Annealed condition. (b) At 0.2% offset.

structure and resultant tensile properties of HIP'ed Ti-6Al-6V-2Sn were evaluated for thick-section ring compacts. Specimen blanks on the order of 3/4 to 1⅛ in. thick were cut from the HIP rings before heat treatment. Screening tests to determine the effects of various annealing treatments and solution treatment-plus-age cycles were performed (Table 11). The effects of different heat treatments on the tensile properties of alpha-plus-beta HIP and beta

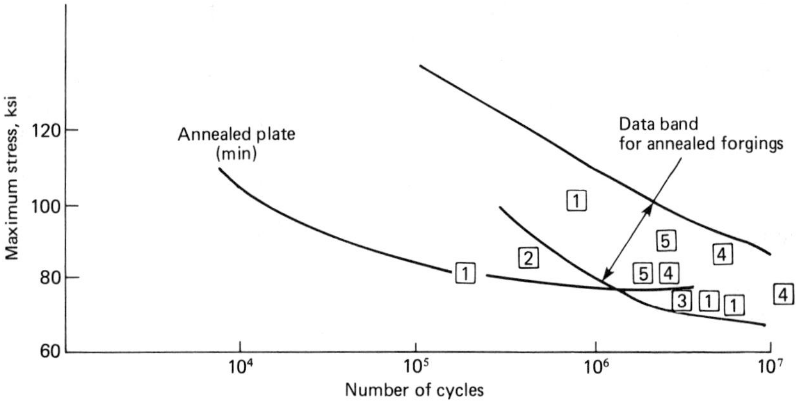

[1] HIP run #1, as machined. [2] HIP run #2, vacuum annealed for 2 h at 1300°F. [3] HIP run #2, vacuum annealed for 16 h at 1300°F. [4] HIP run #2, vacuum annealed for 24 h at 1300°F. [5] HIP run #4, vacuum annealed for 24 h at 1300°F.

Fig. 38. Fatigue endurance of Ti-6Al-6V-2Sn powder consolidated by HIP at 1650°F (Ref 36)

Table 11. Heat treatment conditions for P/M alloy Ti-6Al-6V-2Sn (Ref 37)

Code	Designation	Treatment
AR	As-received	None (as-HIP)
MA	Mill annealed	700°C/2 h/AC
DA	Duplex annealed	900°C/2 h/AC + 785°C/2 h/AC
STA-870/600	Solution treated and aged	870°C/1 h/WQ + 600°C/4 h/AC
STA-870/650	Solution treated and aged	870°C/1 h/WQ + 650°C/4 h/AC
STA-870/700	Solution treated and aged	870°C/1 h/WQ + 700°C/4 h/AC

HIP materials are given in Table 12. Mill-annealed and duplex-annealed properties are equivalent to as-HIP'ed properties except for a reduction in yield strength for mill-annealed alpha-plus-beta HIP material and a reduction in

Table 12. Effect of heat treatment on tensile properties of HIP Ti-6Al-6V-2Sn (Ref 37)

Form	Condition(a)	Yield strength, ksi	Tensile strength, ksi	Total elongation, %	Reduction in area, %
$\alpha + \beta$ HIP(b)	AR	...137	150	24.1	30.0
	MA	...136	145	22.0	39.0
	DA	...140	152	20.2	33.4
	STA-870/700	...149	159	17.1	26.5
β HIP(c)	AR	...131	145	15.1	20.8
	DA	...132	146	10.8	15.6
	STA-870/700	...156	167	8.0	9.8
	STA-870/650	...165	175	3.6	7.0
	STA-870/600	...177	185	1.0	3.4
AMS Specification 4978A	Forged plus annealed	...140	150	8-10	20

(a) Refer to Table 11 for heat treatment conditions. (b) 1675°F/15 ksi/3 h. (c) 1800°F/15 ksi/3 h.

ductility for duplex-annealed beta HIP material. In comparison to wrought material specifications, the yield strength of annealed alpha-plus-beta HIP Ti-6Al-6V-2Sn is near the minimum while total elongation is far better than specification minimums.[38]

Mechanical Properties of Blended Elemental Powder

Blended elemental titanium alloy powder offers a method of making products at a cost much lower than those of prealloyed powder methods due to the low cost of the starting powders. In this system, elemental titanium in the form of sponge fines is combined with elemental alloying additions or master alloys. Consolidation is achieved by blending the powder mix, cold pressing in a die and then sintering. The tensile properties of these materials offer an attractive combination of strength and ductility. However, fatigue properties are substantially lower than those of wrought materials. The poor fatigue behavior is believed to be associated with the presence of pores, chemical inhomogeneities and contaminants, particularly sodium and chlorine.[39,40] Thus, the lower-cost process gives a lower-quality end product, and, to date, parts made from elemental powders have not been considered for critical parts.

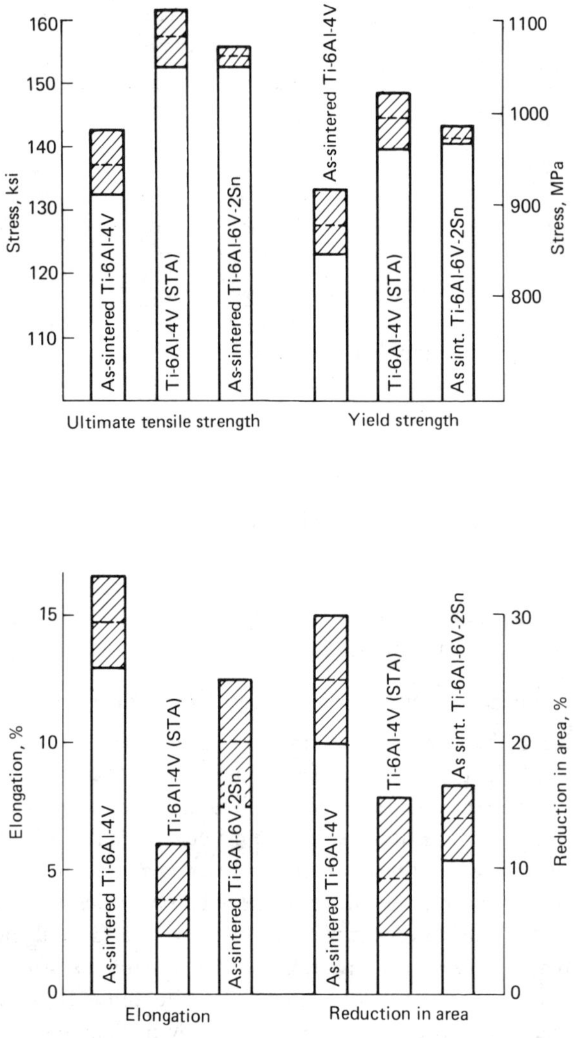

Fig. 39. Typical tensile properties of blended elemental P/M alloys Ti-6Al-4V (as-sintered), Ti-6Al-4V (STA) and Ti-6Al-6V-2Sn (as-sintered) (Ref 41). Shaded areas represent ranges observed.

Table 13. Impact properties of titanium alloys produced from blended elemental powders (Ref 41)

Material	−40°F Charpy V-notch impact, ft·lb Average	Standard deviation
Ti-6Al-4V:		
As-sintered	5.45	0.694
STA(a)	4.90	0.397
Ti-6Al-6V-2Sn, as-sintered	5.39	0.412

(a) Solution treated for 30 min at 1700°F and quenched, then aged for 6 h at 975°F.

The inherent problem of chlorides in the titanium sponge fines remains a significant deterrent to the use of elemental powders.

Eloff *et al*[41] recently reported the results of a study of three titanium alloys made by the blended elemental powder approach. This study resulted in some of the best mechanical properties developed to date for blended elemental powder products. The alloys studied were Ti-6Al-4V (Std), Ti-6Al-6V-2Sn and Ti-6Al-2Sn-4Zr-2Mo. Ti-6Al-4V (Std) and Ti-6Al-6V-2Sn were selected as candidate alloys for a missile component. Emphasis was placed on tensile strength and −40°F impact strength. Tensile properties are presented in bar graphs in Fig. 39, and impact strengths are listed in Table 13. Although the heat treated Ti-6Al-4V (Std) alloy developed the highest strength, the as-sintered Ti-6Al-6V-2Sn alloy was considered the preferred choice because it had the better combination of strength and ductility.

The short-time tensile behavior of Ti-6Al-2Sn-4Zr-2Mo at 1200°F was of interest for a special missile application. Figure 40 gives the room-temperature and 1200°F properties in the as-sintered condition. The room-temperature tensile properties were somewhat superior to those of the Ti-6Al-4V alloy. However, the ductility developed in the 1200°F test condition was unexplainably low.

P/M Titanium Applications

P/M titanium is newer and not as far advanced as superalloys in aerospace applications. Intensive efforts are cur-

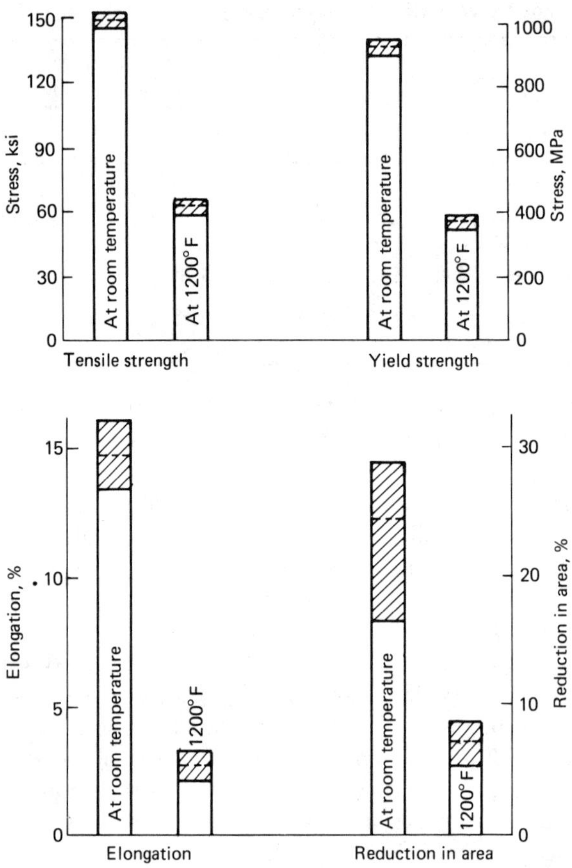

Fig. 40. Tensile properties of blended elemental P/M alloy Ti-6Al-2Sn-4Zr-2Mo at room temperature and at 1200°F (650°C) (Ref 41). Shaded areas represent ranges observed.

rently in progress to develop the ceramic mold shape process for producing near-net engine and airframe hardware from prealloyed powder. Initial studies were conducted using relatively small parts such as the fuselage brace shown in Fig. 41.[42] Current work is aimed at producing large and/or complex airframe and engine parts.[6] A wide variety of parts is being considered for the process. Table 14 lists the parts included in the program and their

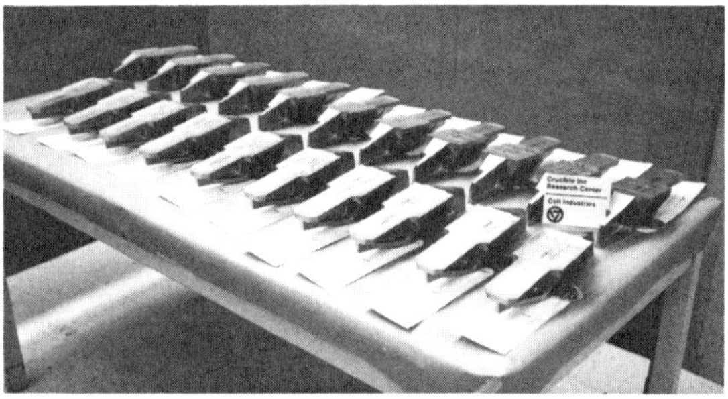

Fig. 41. Pilot production lot of fuselage braces made from Ti-6Al-6V-2Sn powder by the ceramic mold process

Table 14. Trial P/M titanium alloy airframe and engine parts and weight comparison (Ref 6)

Part	Part weight, lb Forging billet	HIP	Final part
Boeing 747 walking beam	55	30	21
General Dynamics F-16 pivot shaft	148	53	32
General Electric TF34 compressor spool	147	65	15
McDonnell Aircraft F-15 drop-out link	115	56	14
McDonnell Aircraft F-15 keel splice	6	0.8	0.4
Northrop F-18 arrestor hook	181	55	28
Pratt & Whitney Aircraft F-100 fan disk	120	65	27
Williams International F-107 compressor rotor	32	6.2	3.6

weights. Figure 42 shows trial P/M titanium alloy airframe parts. Figure 42(a) shows a horizontal stabilizer pivot shaft for the F-16 fighter. This part was produced from Ti-6Al-4V (ELI) powder. The arrestor hook support fitting (Fig. 42b) used on the F-18 fighter is one of the largest P/M titanium alloy parts made to date. It was also made from Ti-6Al-4V (ELI) powder. The Boeing 747 walking beam support fitting (Fig. 42c) and the McDonnell Aircraft

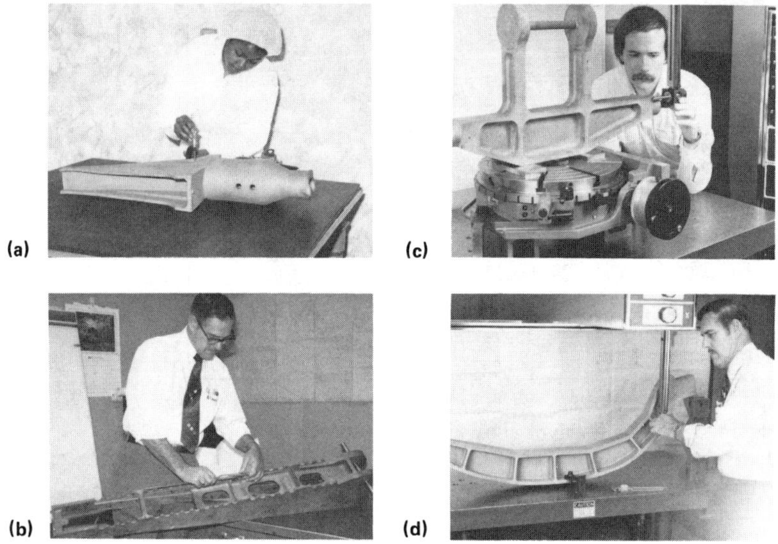

(a) Horizontal stabilizer pivot shaft. (b) Arrestor hook support fitting. (c) Walking beam support fitting. (d) Dropout link.

Fig. 42. P/M titanium alloy airframe shapes made by the ceramic mold process (Ref 6)

drop-out link (Fig. 42d) for the F-15 fighter were made from Ti-6Al-4V (Std) powder.

Engine parts being considered for P/M titanium alloy production are shown in Fig. 43. The compressor rotor (Fig. 43a) for the Williams International F-107 engine was produced from Ti-6Al-4V (Std) powder, as was the General Electric compressor spool for the TF-34 engine (Fig. 43b). Ti-6Al-2Sn-4Zr-6Mo powder was used to produce the Pratt & Whitney Aircraft third-stage fan disk (Fig. 43c) for the F-100 engine.

In another effort,[36,42] Ti-6Al-6V-2Sn powder has been used to produce a fuselage fitting for the F-14A aircraft (Fig. 41). The same alloy is being used to fabricate an F-14A nacelle frame by electron beam welding of four HIP'ed powder parts. Work is also in progress to develop aerospace hardware by cold pressing and sintering blended elemental titanium and master alloys.[41,43] Reportedly, a

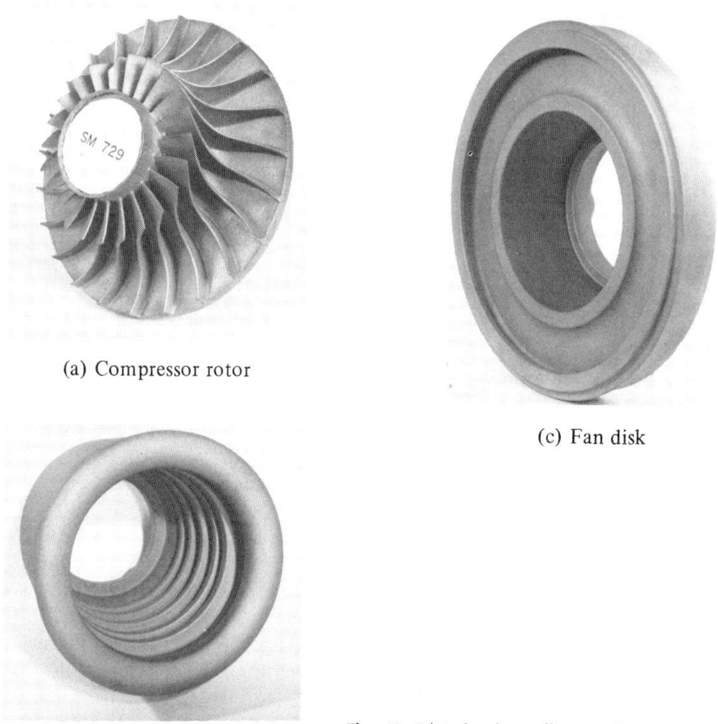

(a) Compressor rotor

(b) Compressor spool

(c) Fan disk

Fig. 43. P/M titanium alloy engine parts made by the ceramic mold process (Ref 6)

near-net-shape compressor stator connecting link for the F-100 engine has been produced by this process. Density above 99% of theoretical and suitable mechanical properties have been achieved.

References

1. J. L. Bartos, "Review of Superalloy Powder Metallurgy Processing for Aircraft Gas Turbine Engines", ASTM STP 672, 1979
2. P. Lowenstein, "Superclean Superalloy Powders", *Powder Metallurgy Superalloys: Aerospace Materials for the 1980's*, MPR Publishing Services, Ltd., England, 1980
3. P. Lowenstein, "Specialty Powders by the Rotating Electrode Process", *Progress in Powder Metallurgy*, Vol 37, MPIF, Princeton, NJ, 1982

4. D. J. Looft and E. C. Van Reuth, "Rapid Solidification Processing: An Overview", *Rapid Solidification Processing*, Claitor's, Baton Rouge, LA, 1980
5. P. R. Holiday, A. R. Cox and R. J. Patterson II, "Rapid Solidification Effects on Alloy Structures", *Rapid Solidification Processing*, Claitor's, Baton Rouge, LA, 1980
6. V. C. Petersen, V. K. Chandhok and C. A. Kelto, "Hot Isostatic Pressing of Large Titanium Shapes", *Powder Metallurgy of Titanium Alloys*, AIME, 1980
7. E. J. Dulis, J. H. Moll, V. K. Chandhok and J. C. Hebeisen, "Progress in P/M Superalloy and Titanium for Aircraft Applications", *The 1980's–Payoff Decade for Advanced Materials*, Vol 25, SAMPE, 1980
8. J. H. Moll, "Outlook for As-HIPed Near-Net Shapes", presented at Workshop on Conservation and Substitution Technology for Critical Materials, Nashville, TN, June 15-17, 1981 (to be published by National Bureau of Standards)
9. Container for Hot Consolidating Powder, U.S. Patent 4,142,888, March 6, 1979
10. J. R. Lizenby, W. J. Rozmas, L. J. Barnard and C. A. Kelto, "Fluid Die Press Consolidation", *Powder Metallurgy Superalloys: Aerospace Materials for the 1980's*, MPR Publishing Services, Shrewsbury, England, 1980
11. Hot Isostatic Pressing Using Vitreous Container, U.S. Patent 3,622,313, November 23, 1971; U.S. Patent Re-issue RE 28,301, January 14, 1975
12. Powder Metallurgy, U.S. Patent 4,227,927, October 14, 1980
13. Process for Compacting Metal Powders, U.S. Patent 3,704,508, December 5, 1972
14. J. D. Buzzanell and L. W. Lherbier, "Processing Effects on the Properties of P/M René 95 Near-Net Shapes", *Superalloys 1980*, ASM, Metals Park, OH, 1980
15. M. M. Allen, R. L. Athey and J. B. Moore, "Nickel-Base Superalloy Powder Metallurgy–State of the Art", *Progress in Powder Metallurgy*, 1975, Vol 31, MPIF, Princeton, NJ, 1975
16. J. E. Coyne, W. H. Couts, C. C. Chen and R. P. Roehm, "Superalloy Turbine Components–Which Is the Superior Manufacturing Process: As-HIP, HIP + Isoforge or GATORIZING™ of Extrusion Consolidated Billet", *Powder Metallurgy Superalloys–Aerospace Materials for the 1980's*, Vol 1, MPR Publishing Services, Shrewsbury, England, 1980
17. E. J. Dulis and J. H. Moll, "Properties of HIP Consolidated Superalloy and Titanium Alloys", *Rapid Solidification Processing*, Claitor's, Baton Rouge, LA, 1978

18. R. H. Van Stone, F. J. Rizzo and J. F. Radavich, "Microstructures of Gas Atomized René 95 Powder", *Rapid Solidification Processing II*, Claitor's, Baton Rouge, LA, 1980
19. R. Mehrabian, "Relationship of Heat Flow to Microstructure in Rapid Solidification Processing", *Rapid Solidification Processing*, Claitor's, Baton Rouge, LA, 1978
20. M. Cohen, B. H. Kear and R. Mehrabian, "Rapid Solidification Processing–An Overlook", *Rapid Solidification Processing II*, Claitor's, Baton Rouge, LA, 1980
21. Data developed at Crucible Research Center, Pittsburgh, PA
22. J. L. Bartos and P. S. Mathur, "Development of Hot Isostatically Pressed (As-HIP) Powder Metallurgy René 95 Turbine Hardware", *Superalloys–Metallurgy and Manufacture*, AIME, Claitor's, Baton Rouge, LA, 1976
23. J. F. Barker and E. H. VanDerMolen, "Effect of Processing Variables on Powder Metallurgy René 95", *Superalloys–Processing*, AIME, Battelle, Columbus, OH, 1972
24. D. J. Evans and G. M. Judd, "Influence of Atomization Stock on Properties of Hot Isostatically Pressed, Low-Carbon Astroloy", *Modern Developments in Powder Metallurgy*, Vol 11, MPIF, Princeton, NJ, 1977
25. M. T. Podob, "Effect of Heat Treatment and Slight Chemistry Variations on the Physical Metallurgy of Hot Isostatically Pressed, Low-Carbon Astroloy Powder", *Modern Developments in Powder Metallurgy*, Vol 11, MPIF, Princeton, NJ, 1977
26. D. J. Evans and R. D. Eng, "Development of a High Strength Hot-Isostatically-Pressed Disk Alloy, MERL 76", *Modern Developments in Powder Metallurgy*, Vol 14, MPIF, Princeton, NJ, 1980
27. R. L. Dreshfield and R. V. Miner, Jr., "Application of Superalloy Powder Metallurgy for Aircraft Engines", *P/M Superalloy Technology and Applications*, 1980 International Powder Metallurgy Conference
28. J. H. Moll, J. H. Schwertz and V. K. Chandhok, "P/M Dual-Property Wheels for Small Engines", *Progress in Powder Metallurgy*, Vol 37, MPIF, Princeton, NJ, 1982
29. C. S. Kortovich and J. M. Marder, "Development of Materials and Process Technology for Dual Alloy Disks", NASA CR-165224, October 1981
30. G. Friedman, "The Powder Metallurgy of High-Strength Titanium Alloys", *Metal Progress*, ASM, March 1975
31. V. C. Petersen and V. K. Chandhok, Interim Report, Contract F33615-77-C-5005, June 1979
32. G. W. Kuhlman and F. R. Billman, "Selecting Process-

ing Options for High-Fracture-Toughness Titanium Airframe Forgings", *Metal Progress*, March 1977
33. C. S. Barker and P. Nagy, "Application of HIP Titanium Ti-6Al-4V Powder to the Radial Compressor for the Cruise Missile Engine", *Powder Metallurgy of Titanium Alloys*, AIME, 1980
34. J. H. Schwertz, V. K. Chandhok, V. C. Petersen and V. R. Thompson, "Consolidation of Titanium Powder to Near-Net Shapes", AFML-TR-78-41, May 1978
35. G. R. Chanani *et al*, "Application of Titanium P/M for Manufacture of a Large and Complex Naval Aircraft Component", *Powder Metallurgy of Titanium Alloys*, AIME, 1980
36. R. H. Witt and J. Magnuson, "Flight Qualification of Titanium F-14A Airframe Components Manufactured by Hot Isostatic Pressing", Final Report, Contract N00019-76-C-0143, Grumman Aerospace, June 1977
37. D. B. Dawson and G. Ulitchny, "Characterization of the Tensile Properties of Hot Isostatically Pressed Ti-6Al-6V-2Sn", *Powder Metallurgy of Titanium Alloys*, AIME, 1980
38. W. E. Brown, Jr., *Aerospace Structural Metals Handbook*, Vol 4, Code 3715, Battelle, 1979
39. Y. Mahajan, D. Eylon, R. Bacon and F. H. Froes, "Microstructure Property Correlation on Cold Pressed and Sintered Elemental Ti-6Al-4V Powder Compacts", *Powder Metallurgy of Titanium Alloys*, AIME, 1980
40. R. R. Boyer, J. E. Magnuson and J. W. Tripp, "Characterization of Pressed and Sintered Ti-6Al-4V Powders", *Powder Metallurgy of Titanium Alloys*, AIME, 1980
41. P. C. Eloff *et al*, "Properties of P/M Titanium Alloys Produced from Elemental Blends: The Current Status of Development", *Progress in Powder Metallurgy 1981*, MPIF, 1982
42. R. H. Witt and W. T. Highberger, "Hot-Isostatic Pressing of Near-Net Titanium Structural Parts", *Powder Metallurgy of Titanium Alloys*, AIME, 1980
43. P. J. Anderson and P. C. Eloff, "Development of Higher Performance Blended Elemental Powder Metallurgy Titanium Alloys", *Powder Metallurgy of Titanium Alloys*, AIME, 1980
44. J. W. Eggar and R. J. Siddall, "A Review of the Techniques Available for the Production of Superalloy Powders", *P/M Superalloy Technology and Applications*, Seminar Preprints, 1980 International Powder Metallurgy Conference
45. G. König, W. Track and W. Betz, "Assessment of LCF Behavior of a P/M Nickel-Base Alloy for Turbine Disc Applications", *Metal Powder Report*, Vol 38, No. 1, January 1983

Index

NOTE: This index deals almost exclusively with powder metallurgy materials; therefore, the designation P/M and the words powder or powder metallurgy are used sparingly and only for clarity. The form of any material that is not powder and is used for comparison has been noted. The (F) or (T) following an entry indicates that information is in a Figure or a Table.

A

Accelerator. *See* Machine gun accelerator
Aerospace applications
advantages 247-248
current status 248
process (F) 248-258
superalloys (F,T) 258-279
titanium alloys (F,T) 279-295
Aft shafts, of nickel superalloys (T) . . 269-271
Aging
control of nitrogen 205
effect of carbon pickup (F) 206-208
effect of sintering atmosphere in iron compacts 200
effect on coercive force in iron compacts (F) 203-204
effect on induction of iron compacts (F) 200-201
effect on maximum permeability of iron compacts (F) 202-203
effect on remanent magnetization of iron compacts (F) 201-202
Airframe parts
made by ceramic mold process (F) 294
of titanium alloys (F,T) 293-294
Alloying
during sintering 38-41
examples of during sintering (F) 38-39
Aluminum alloys
applications 224-233
compacting (F) 216-218
composition of Alcoa aluminum premixes (T) 212
compressibility (F) 216-217
corrosion resistance (T) 214-215
electrical and thermal conductivity (T) 215-217
fatigue (F) 214
finishing (F) 221-222
for appliances (F) 229-230
for automotive components (F) 227-228
for business machines (F) 224-227
for power tools (F) 229-231
heat treatment 220
joining (F) 222-224
light weight (F) 215-216
machining (F) 220-221
material characteristics 211-212
properties (T) 213
sintering 218-219
sizing, coining and cold forming . . . 219-220
tempers for sintered alloys (T) 213
Aluminum alloys, specific types
201AB
composition (T) 212
corrosion resistance (T) 215
fatigue (F) 214
properties (T) 213
202AB
composition (T) 212
fatigue (F) 214
properties (T) 213
601AB
composition (T) 212
corrosion resistance (T) 215
fatigue (F) 214
properties (T) 213
2014 (wrought)
corrosion resistance (T) 215
6061 (wrought)
corrosion resistance (T) 215

Aluminum alloys, wrought
composition (T) 238-239, 241
corrosion resistance 236-237
electrical conductivity (T) 217
fracture toughness/strength relationships
(F) 238-239
grain structure (F) 239, 241
manufacture 242-246
notched fatigue strength (F) 238, 240
physical properties (F) 242-243
process 235
properties 235-236
resistance to atmospheric conditions 237
resistance to exfoliation 236-237
resistance to stress corrosion cracking
(F) 237-238
tensile properties (T) 236-237
thermal conductivity (T) 217
types 235
Aluminum alloys, wrought, specific types
7090
composition (T) 238-239, 241
manufacture (F) 242-246
notched fatigue strength (F) 238, 240
physical properties (T) 242-243
resistance to atmospheric conditions . .237
stress corrosion cracking (F) 237-238
tensile properties (T) 236-237
7091
composition (T) 238-239, 241
grain structure (F) 239, 241
manufacture 242-243
notched fatigue strength (F) 238, 240
physical properties (T) 242-243
resistance to atmospheric conditions . .237
stress corrosion cracking (F) 237-238
tensile properties (T) 236-237
2024
notched fatigue strength (F) 240
7050
composition (T) 241
grain structure (F) 239, 241
notched fatigue strength (F) 238, 240
physical properties (T) 242-243
stress corrosion cracking 237-238
7075
composition (T) 241
notched fatigue strength 238, 240
physical properties (T) 242-243
stress corrosion cracking 237-238
7171
composition (T) 241
7175
composition (T) 241
physical properties (T) 242-243

Anvil pressing 44
Apparent density 17-18
effect of fine particle additions (F) 18-19
measurement of 17
Appliances
of aluminum alloys (F) 229-230
use of self-lubricating porous bronze bearings
(T) 154
Applications 2-3
appliances of aluminum alloys (F) . . 229-230
automotive components of aluminum alloys
(F) 227-228
breakdown of total volume of stainless steels
by market segment for 1973 and 1980
(T) 139
business machines of aluminum alloys
(F) 224-227
of aluminum alloys (F) 224-233
of copper and copper alloys
(F,T) 146, 151-183
of copper and iron-base friction materials
for brakes and clutches (F,T) . . 159-162
of copper-base structural parts
(F) 163-165, 167, 169
of copper infiltrated steel parts (F) . . 177-178
of dual-alloy, dual-property components
(F) 278-279
of dual-property turbine wheels (F) 278-281
of ferrous sintered material 73, 76-77, 82-83
of nickel superalloy multicomponent parts
(F) 255, 276
of nickel superalloy turbine-engine compo-
nents (F) 275-277
of stainless steel porous parts (F) . . . 141-144
of stainless steel structural parts (F) 138-141
of superalloys for aerospace use (F) 275-279
of titanium alloys for aerospace applications
(F) 291-295
power tools of aluminum alloys (F) 229-231
Atmospheres
for sintering (T) 70-71
mixtures of nitrogen and hydrogen gases for
stainless steels 130-131
of dissociated ammonia for stainless
steels 130-131
of hydrogen for stainless steels 132-133
Atmospheric conditions
resistance to in wrought aluminum
alloys 237
Atomization (*see also* particular types, e.g.,
Centrifugal atomization; Gas atomization;
Vacuum atomization; Water atomization)
production of metal powders (F) 8-9
Austenitic stainless steels
effect of tin on corrosion resistance 138

Austenitic stainless steels, specific types
Type 303L
composition (T) 127
hardware parts (F) 140
price in U.S. in 1982 (T) 127
Type 304L
appliance parts (F) 140
composition (T) 127
hardware parts (F) 140
price in U.S. in 1982 (T) 127
Type 316L
composition (T) 127
effect of density on sintering properties (F) 132, 135
effect of lubricant and green density on green strength (F) 129
effect of nitrogen on mechanical properties (F) 132-134
effect of porosity on acoustic response (F) 141, 143
effect of sintering temperature on mechanical properties (F) 133-135
filter parts (F) 144
handle blanks for hunting knife (F) . . . 141
microstructure of acceptable and unacceptable parts (F) 138-139
office-machine parts (F) 140
price in U.S. in 1982 (T) 127
properties after vacuum sintering (T) . . 136
relationship between flow and pressure drop (F) 142
use for porous parts 141
Automotive components
of aluminum alloys (F) 227-229
of ferrous powder (F) 76, 82-83
use of self-lubricating porous bronze bearings (T) 154

B

Bearings
cold forging 112
of ferrous forgings (F) 109, 111-113
Bevels 49-50
Brakes
of copper and iron-base friction materials (T) 160-162
Bridging effect (F) 18
Bronze bearings, porous, self-lubricating (F) 151-158
advantages and disadvantages (T) . 155, 158
in automotive components (T) 154
in business machines (T) 154
in home appliances (T) 154
in industrial equipment (T) 154
in leisure industry (T) 154
MPIF industry standard (T) 154
properties 152, 154
PV factor vs life of (F) 158
shapes 153, 155
Business industry
use of self-lubricating porous bronze bearings (T) 154
Business machines
of aluminum alloys (F) 224-227

C

CAP® (consolidation by atmospheric pressure)
of nickel superalloys and titanium alloys 257
Carbide distributions
in nickel superalloys (F) . . 261-262, 264-267
Carbon brushes
from copper-base flakes 182
Carbon pickup
effect on coercive force in iron-phosphorus compacts (F) 207-208
effect on induction in iron-phosphorus compacts (F) 206-207
effect on permeability in iron-phosphorus compacts (F) 207-208
effect on remanent magnetization in iron-phosphorus compacts (F) 207
Centrifugal atomization
for nickel superalloys and titanium alloys 252-253
Ceramic mold process
for titanium alloy airframe parts (F) 294
Chamfers 50
Chemical decomposition
production of metal powders 9
Clutches
of copper and iron-base friction materials (T) 160-162
Coatings 53
Coefficient of friction
effect of in copper-base friction materials 163
Coercive force
effect of aging in iron compacts (F) 203-204
effect of carbon pickup in iron-phosphorus compacts (F) 207-208
effect on iron-phosphorus alloys (F) 192-194
Coining (F) 51-52
in aluminum alloys 219
Cold forging
of ferrous bearings 112
Cold forming, in aluminum alloys 220
Cold isostatic compaction
of ferrous powders (F) 65-67

Comminution methods
production of metal powders 9
Compaction 21-22, 45
behavior of powder 25-28
double-action (F)................. 22, 24
floating-action (F) 24-25
green density (F) 27-28
green strength 28-29
in rigid dies (F) 22-24
of aluminum alloys 216, 218
of ferrous powders (F) 62-66
of stainless steels (F).............127-128
single-action (F) 22, 24
techniques 22
Composite side gear
of forged 4620 steel (F) 106-108
Composition
of stainless steels (T).................127
of wrought aluminum alloys
(T) 238-239, 241
of wrought I/M aluminum alloys (T) 241
Compressibility (F) 19-21
Consolidation 4-5
Containerization
Kelsey-Hayes fluid die process 257
Kelsey-Hayes glass bag process 257
of nickel superalloys (F) 253-257
Universal-Cyclops CAP® process 257
use of ceramic mold (F) 254, 256
Copper and copper alloys (*see also* Bronze bearings; Graphite bearings)
applications (F,T) 146, 151-183
atomization (F,T) 146-149
effect of density on electrical conductivity
(F) 170-171
electrical parts (F,T) 169-177
electrolysis (F,T) 146-147, 149-150
friction materials (F,T) 158-163
hydrometallurgy (F,T) ... 146-147, 150-151
infiltrated parts (F,T).... 172, 174, 176-182
oxide reduction (F,T) 146-147, 150
production of powders 145-151
properties (T).......................146
raw materials for alloys (T) 151
self-lubricating porous bronze bearings
(F,T) 151-158
structural parts (F)...............164-169
Copper and copper alloys, specific types
70/30 leaded brass (B-155)
mechanical properties (F) 166
80/20 leaded brass (B-129)
mechanical properties (F) 166
90Cu-10Sn
microstructure of bronze parts after
time in hot zone (F) 152
brass, wrought
electrical conductivity (T)...........217
thermal conductivity (T) 217
bronze, wrought
electrical conductivity (T)...........217
thermal conductivity (T) 217
Copper-infiltrated parts
(F,T) 172, 174, 176-182
Copper-infiltrated steel parts...... 172, 174
applications (F) 177-178
developments.......................176
effect of infiltration (T) 178
effect of infiltration on transverse rupture
strength (F) 179
microstructure (F) 179, 182
MPIF standard (T)...............179-181
sintration (F) 177-178
Copper infiltration
effects of (F,T) 178-179
Copper, wrought
electrical conductivity (T).............217
thermal conductivity (T) 217
Corner reliefs 50
Corrosion resistance
effect of tin in austenitic stainless steels..138
of aluminum alloys (T) 214-215
of bronze structural parts 165, 167
of stainless steels 135, 137-138
of wrought aluminum alloys 236-237
Cost
of stainless steels in 1982 in U.S. (T) ...127

D

DC properties
effect of phosphorus in iron alloys
(F) 186-188
Dendrite-arm spacing
in nickel superalloys (F) .. 259-260, 262-263
Density
effect on saturation in iron alloys (F)....188
effect on stainless steels (F) .. 132, 135, 143
in iron-phosphorus compact (F) 195-196
Design
of powder metal parts 47-51
Die compaction
of ferrous powders (F) 63-65
Differential gear set
for army truck (F) 103-106
of forged 4600 steel 104-105
Diffusion
during sintering (F)................37-38
Direct reduction of oxides
Hoeganaes process (F) 56-59
production of ferrous powders 56-59

Pyron process (F) 56-59
Disks (*see also* Turbine disks)
mechanical properties of IN-100 (T) 273
Distribution
of parts from specialty alloys in 1977 (F) 123, 125
of stainless steel parts in 1979 (F) 126
Double-action compaction (F) 22, 24
Dual-alloy, dual-property components
development of (F) 278-279
Dual-property turbine wheels
development of (F) 278-281

E

Elastic modulus
of ferrous sintered materials (F) 75
Electrical conductivity
of aluminum alloys (T) 215-217
of wrought aluminum alloys, brass, bronze, iron and copper (T) 217
Electrical parts, copper-base 169-170
Electrical parts, copper-base, specific types
GlidCop
applications (F) 170, 174, 176
properties (F) 170, 172-174
welding conditions for various applications (T) 175
Al-20 GlidCop
effect of elevated temperature on mechanical properties (F) 173
elevated stress-rupture properties (F) . .173
Al-60 GlidCop
elevated stress-rupture properties (F) . .174
Electrolysis
of copper and copper alloys (F,T) 146-147, 149-150
Engine parts
of titanium alloys (F,T) 293-295
Exfoliation
resistance to in wrought aluminum alloys 236-237
Extrusion
plus hot working in tool steels 118

F

Fatigue
of aluminum alloys (F) 214
of 4640 steel (F) 102-103
Fatigue properties
of titanium alloys (F) 284-287
Fatigue strength
in wrought aluminum alloys (F) . . . 238, 240
in wrought I/M aluminum alloys (F) 238, 240
Ferritic stainless steels, specific types
Type 434L
composition (T) 127
price in U.S. in 1982 (T) 127
properties after sintering in dissociated ammonia (T) 136
Ferrous powder metallurgy (*see also* Iron; Iron alloys, specific types)
advantages 55-56
compaction and sintering technology (F,T) 62-83
composition of materials (F) 77
density designations and ranges (T) 77
forging technology 86-113
friction materials (T) 159, 162
high-density metallurgy 83-86
mechanical properties (T) 78-81
production of 56-62
superalloys 119
tool steels 113-119
Ferrous superalloys 119
Fillers
for copper-base friction materials 163
Finishing
of aluminum alloys (F) 221-222
Finishing, decorative
of copper-base flakes 182
Fins. *See* Guide fins
Flakes, copper-base
as carbon brushes 182
as decorative finishing 182
Floating-action compaction (F) 24-25
Flow rate 18-19
Forging
applications of ferrous forging 99-113
cost compared with conventional forging (T) 104
effect of forging variables on mechanical properties, (F) 97-99
of bearings (F) 109, 111-113
of composite side gear (F) 106-108
of differential gear set for army truck (F) 103-106
of ferrous powders 86-112
of guide fins (F) 106-107, 109
of machine gun accelerator (F) 99-103
preform design (F) 87-95
preform production 86-87
process 95-97
Fracture toughness
strength relationships in wrought aluminum alloys (F) 238-239
Friction materials 158-159
applications (T) 159-162
coefficient of friction 163

Friction materials *(continued)*
fillers 163
for brakes (T) 160
for clutches (T) 160
functions and component used (T) 162
lubricant 163
matrix 163
of copper-base materials (F,T) 159, 161-162
of iron-base materials (T) 159, 162
performance characteristics (T) 162
processing technology 159-160
thermal stability 163
Future
of stainless steel parts 142, 144

G

Gas atomization
microstructure after, in nickel superalloys (F) 259-261
of copper and copper alloys (F,T) . . 146-149
of nickel superalloys (F) 249, 251
GATORIZING™
of nickel superalloys and titanium alloys 257
Gear pumps
application of ferrous powder 73
material requirements (T) 82
Grain size
effect on magnetic properties of iron-phosphorus alloys (F) 189-192
Grain structure
of wrought aluminum alloys (F) . . . 239, 241
of wrought I/M aluminum alloys (F) 239, 241
Graphite bearings, nonporous 155
Green density (F) 27-28
vs green strength (F) 29
Green strength 28-29
vs green density (F) 29
Guide fins
characteristics of oil-quenched and coin-tempered fins (T) 111
for Navy guided projectile (F) .106-107, 109
of forged 4640 steel (F) 106-107, 109

H

Hardness
relationship between apparent hardness and particle hardness (F) 76
Heat treatment 54
effect on tensile properties in titanium alloys (T) 287, 289
of aluminum alloys 220
of titanium alloys (T) 286-288
High-density ferrous powder metallurgy .83-86
composition and properties of infiltrated steels (T) .84-85
infiltration 84-86
High-temperature sintering. *See* Sintering, high-temperature
History 1-2
Hoeganaes process
composition and properties of Hoeganaes sponge iron powder (T) 59
for production of ferrous powders (F) . 56-59
Home appliances. *See* Appliances
Hot isostatic pressing
of nickel superalloys and titanium alloys 257
plus hot working in tool steels 118
Hot working
with extrusion in tool steels 118
with hot isostatic pressing in tool steels . .118
Hydraulic presses
advantages and disadvantages 44
Hydrometallurgy
of copper and copper alloys (F,T) 146-147, 150-151

I

Impact properties
of titanium alloys (T) 291
Induction
effect of aging in iron compacts (F) 200-201
effect of carbon pickup in iron-phosphorus compacts (F) 206-207
Industrial equipment
use of self-lubricating porous bronze bearings (T) 154
Infiltrted steels (*see also* Copper-infiltrated steel parts)
composition and properties (T) 84-85
Infiltration
in high-density ferrous powder metallurgy 84-86
Interparticle bonding
in sintering 33
Iron
composition and properties of Hoeganaes sponge iron powder (T) 59
composition and properties of Pyron iron powder (T) 59
compressibility (F) 217
magnetization curves (F) 187-188
predictor equation for resistivity (F) 200
Iron, wrought
electrical conductivity (T) 217
thermal conductivity (T) 217

Iron alloys, specific types
Fe-P
aging 204-208
control of nitrogen in sintering 205
effect of carbon pickup (F) 206-208
density (F) 195-196
effect of coercive force (F) 192-194, 197-198
effect of grain size (F) 189-192
effect of oxygen content (F) 192
effect of phosphorus on DC properties (F) 186-188
effect of porosity on saturation (F) 187
effect of purity (F) 189-190
effect of sintering temperature on oxygen content (F) 189
full-density parts 188-193
magnetization curves (F) 187
maximum permeability (F) 197
porous compacts 193-199
regression analysis 195-196
residual induction (F) 196-197
resistivity 199-200
Fe-0.45P
comparison of coercive force as a function of grain intercept (F) 197-198
comparison of effect of density on induction (F) 195-196
comparison of maximum permeability as function of grain intercept (F) 197
comparison of maximum permeability with predictor equation (F) 193
effect of oxygen and grain intercept on residual induction (F) 191
predictor equation for resistivity as a function of density (F) 200
residual induction as a function of pore area (F) 196
Iron compacts
effect of aging on coercive force (F) 203-204
effect of agingon induction (F) 200-201
effect of aging on maximum permeability (F) 202-203
effect of aging on remanent magnetization (F) 201-202
effect of sintering atmosphere on aging . . 200
Isostatic pressing 44
advantages 65
dry bag and wet bag methods (F) 66-67
of ferrous parts 65-67

J

Joining (F) 52-53
of aluminum alloys (F) 222-224

K

Knurls 49

L

Leisure industry
use of self-lubricating porous bronze bearings (T) 154
Length-to-height ratio 48-49
Lubricant
for copper-base friction materials 163
for copper-base structural parts 164
for stainless steels 127-129

M

Machine gun accelerator
application of forging (F) 99-103
of forged steel (F) 100-103
Machining 52-53
of aluminum alloys (F) 220-221
Manufacture
of wrought aluminum alloys (F) 242-246
Manufacturing processes 3-4
Martensitic stainless steels, specific types
Type 410L
composition (T) 127
price in U.S. in 1982 (T) 127
properties after sintering (T) 135-136
Matrix
for copper-base friction materials 163
Mechanical presses
advantages and disadvantages 44
Mechanical properties
development of during sintering 30-32
effect of forging variables (F) 97-99
effect of level of upsetting (F) 97, 99
effect of porosity (F) 31, 97-98
effect of sintering variables (F) 97, 99
of blended elemental titanium alloy powder (F,T) 289-291
of ferrous sintered materials 73
of ferrous materials (T) 78-81
of iron compacts (F) 74
of nickel superalloys (F,T) 268-275
of prealloyed titanium powder (F,T) 283-289
Mechanisms
of sintering 36-38
Metal powders
apparent density 17-18
atomization (F) 8-9
chemical decomposition 9
comminution methods 9
compaction (F) 21-29

Metal powders *(continued)*
compressibility (F) 19-21
development of mechanical properties 30-32
development of microstructure 32-36
flow rate 18-19
particle shape (F) 14-17
particle size (F) 11-14
particle size distribution (F) 11-15
production (T) 7-10
sintering 29-41
thermal decomposition 9
Microstructure
development of during sintering 32-36
interparticle bonding 33
neck growth (F) 33
of copper-infiltrated steel parts (T) 179, 182
of titanium alloys 282-283
pore channel closure 33-34
pore coarsening 35-36
pore rounding (F) 34-35
pore shrinkage 34-35
MPIF standard
for bronze (T) 154, 156-157
for copper-infiltrated steel (T) 180-181
Multicomponent parts
use of nickel superalloys (F) 255, 276

N

Neck growth in sintering (F) 33
Nickel superalloys
aerospace applications (F) 275-279
carbide networks 261-262, 264
centrifugal atomization 252-253
composition (T) 258
consolidation 257
containers for hardware (F) 253-257
dendrite-arm spacing (F) . 259-260, 262-263
finishing 257-258
forging 258
gas atomization (F) 249, 251
mechanical properties (F,T) 268-275
microstructure after gas atomization (F) 259-261
plasma rotating electrode process (PREP) (F) 251-253
powder characteristics (F) 258-267
screening for oversize particles 253
vacuum atomization (F) . 249, 252, 258-259
Nickel superalloys, specific types
Astroloy, low-carbon
composition (T) 258
mechanical properties (T) . . . 269, 271-272
IN-100
carbide distributions (F) 264-265
carbide networks 261-262, 264
composition (T) 258
effect of thermal exposure on carbides (F) 265-266
mechanical properties 271, 273
mechanical properties of disk (T) 273
microstructure after gas atomization (F) 260
IN-792
carbide distributions (F) 264-265
effect of thermal exposure on carbides (F) 265, 267
MERL 76
composition (T) 258
effect of section size on tensile properties (T) 275
fatigue life (F) 276
mechanical properties 273-274
stress-rupture properties (F) 275
tensile strength vs temperature (F) 274
René 95
composition (T) 258
mechanical properties 268-271
mechanical properties of aft shafts (F) 269-271
mechanical properties of turbine disks (T) 269
mechanical properties of turbine wheels (T) 268
variation of dendrite-arm spacing with powder size (F) 262
Nitrogen
control of in iron-phosphorus compacts . . 205
effect on stainless steels during sintering (F) 130-134

O

Outdoor garden equipment
application of ferrous powder 73, 76
Oxide reduction
of copper and copper alloys (F,T) 146-147, 150
Oxygen content
effect on maximum permeability in iron-phosphorus alloys (F) 192
effect on remanent magnetization in iron-phosphorus alloys (F) 190-191
effect of sintering temperature in iron-phosphorus alloys (F) 189

P

Particle shape 14-17
types of (F) 16
Particle size (F) 11-14

Particle size distribution (F) 11-14
Permeability
effect of aging in iron compacts (F) 202-203
effect of carbon pickup in iron-phosphorus compacts (F) 207-208
in iron-phosphorus alloys (F) . . 192-193, 197
Phase diagram, rhenium-tungsten (F) 40
Phosphorus
effect on properties of iron alloys . . . 185-209
Phosphorus alloys, specific types. *See* Iron alloys, specific types
Physical properties
of wrought aluminum alloys (T) 242-243
of wrought I/M aluminum alloys (T) 242-243
Pistons, shock-absorber
of aluminum alloys (F) 227-228
Plasma rotating electrode process (PREP)
for nickel superalloys and titanium alloys (F) 251-253
for preparation of titanium alloys 279, 282-283
microstructure of titanium alloys . . . 282-283
Plating 53
Pore channel closure
in sintering 33-34
Pore coarsening
in sintering 35-36
Pore rounding
in sintering (F) 34-35
Pore shrinkage
in sintering 34-35
Porosity
effect in stainless steels (F) 141-144
effect on mechanical properties (F) . . . 97-98
effect on saturation in iron alloys (F) 187-188
of ferrous parts after sintering 69
Porous parts
of stainless steels (F) 141-144
Powder preparation 4
Power tools
of aluminum alloys (F) 229-231
Preform design (F) 87-95
defects (F) 91-94
re-pressing (F) 88-90
Preform production 86-87
Pressing equipment 43-44
Purity
effect on magnetic properties in iron-phosphorus alloys (F) 189-190
PV factor
vs life of bronze bearings (F) 158
Pyron process
composition and properties of Pyron iron powder (T) 59
for production of ferrous powders (F) . 56-59

R

Raw materials
for copper alloys (T) 151
Reduction with carbon. *See* Hoeganaes process
Reduction with hydrogen gas. *See* Pyron process
Regression analysis
for iron-phosphorus compacts 195-196
Remanent magnetization
effect of aging in iron compacts (F) 201-202
effect of carbon pickup in iron-phosphorus compacts (F) 207
effect of oxygen content in iron-phosphorus alloys (F) 190-191
Re-pressing (F) 51-53, 88-90
Residual induction
in iron-phosphorus compacts (F) . . . 196-197
Resistivity
in iron-phosphorus compacts (F) . . . 199-200
Rhenium-tungsten
phase diagram (F) 40
Rigid die compaction (F) 22-24
Rotary presses 44
Rotating electrode process (REP)
for preparation of titanium alloys 279, 282-283
microstructure of titanium alloys . . . 282-283

S

Sales
of stainless steels, 1962-1982 (F) . . 123, 125
Saturation
effect of density in iron alloys (F) 188
effect of porosity in iron alloys (F) . 187-188
Secondary processing 51-54
Sensor housings
of aluminum alloys (F) 228-229
Serrations 49
Shapes 47-48
that cannot be molded 50
Sieve analysis
of stainless steels (T) 127
Single-action compaction (F) 22, 24
Sintering 5
applications after sintering of ferrous parts 73
development of mechanical properties 30-32
development of microstructure 32-36
effect of atmosphere on aging in iron compacts 200
effect of tempering on oxygen content in iron (F) 189
effect on mechanical properties (F) . . 97, 99

Sintering *(continued)*
for alloying (F) 38-41
for copper-base structural parts 164-165
hydrogen sintered stainless steels . . . 132-133
interparticle bonding 33
mechanisms 36-38
neck growth (F) 33
of aluminum alloys 218-219
of ferrous parts (F) 66, 68-69, 72
of stainless steels 129-135
pore channel closure 33-34
pore coarsening 35-36
pore rounding (F) 34-35
pore shrinkage 34-35
properties after sintering of ferrous parts 69, 73
use of dissociated ammonia for atmosphere for stainless steels 130-131
vacuum sintering of stainless steels 132-134
Sintering, high-temperature
of tool steels 118-119
Sintration (F) 176-177
Sizes of parts 47-48
Sizing
in aluminum alloys 219
Specialty steels 112-119
Specifications
for composition of alloys in copper-base structural parts 165
Splines 49
Stainless steels (*see also* Austenitic stainless steels; Ferritic stainless steels; Martensitic stainless steels)
applications 138-141
breakdown of total volume of applications by market segment for 1973 and 1980 (T) 139
compaction (F) 127-128
composition (T) 125-127
corrosion resistance 135, 137-138
distribution of parts (F) 123-124, 126
effect of density on tensile and yield strength (F) 143
future 142, 144
lubricant 127-129
porous parts (F) 141-144
preparation 123, 125
prices of in U.S. in 1982 (F) 127
sales, 1962-1982 (F) 123, 125
sieve analysis (T) 125, 127
sintering (F) 129-135
structural parts (F) 138-141
Staking (F) 52-53
Steam treating 54
Steels, specific types
4600
differential gear set by forging (F) 103-106
4620
composite side gear by forging (F) 106-108
4640
fatigue (F) 102-103
for guide fins for Navy guided projectile (F) 106-107, 109
for machine gun accelerator by forging (F) 99-103
tensile properties compared with wrought 4640 steel (F) 102
Cu steel (5Cu-0.8C)
strength/density curves (F) 169
Strength
relationship with fracture toughness in wrought aluminum alloys (F) . . 238-239
Stress corrosion cracking
resistance to in wrought aluminum alloys (F) 237-238
Structural parts
of stainless steels (F) 138-141
Structural parts, copper-base
applications 163
processing 164
Structural parts, copper-base, specific types
brass parts
applications (F) 164-165, 167
compositions available 164
lubricant 164
sintering 164-165
specifications for composition of alloys 165
bronze parts
applications (F) 167
corrosion and wear resistance . . . 165, 167
90/10 bronze compacts
effect of density on strength (F) 168
Sn bronze parts 167
Cu-Ni powders
applications 167, 169
Ni-Ag compacts (65Cu-18Ni-17Zn)
applications (F) 164, 167
lubricant 164
sintering 164-165
specifications for composition of alloys 165
Surface finish 51

T

Tempers
for aluminum alloys (T) 213
Tensile properties
effect of heat treatment in titanium alloys (T) 287, 289

of 4640 steel compared with wrought 4640 steel (F) 102
of titanium alloys (F,T) 284, 287, 290
of wrought aluminum alloys (T) 236-237

Thermal conductivity
of aluminum alloys (T) 215-217
of wrought aluminum alloys, brass, bronze, iron and copper (T) 217

Thermal decomposition
production of metal powders 9

Thermal stability
for copper-base friction materials 163

Tin
effect on corrosion resistance in austenitic stainless steels 138

Titanium alloys
aerospace applications 291-295
centrifugal atomization 252-253
consolidation 257
containers for hardware (F) 253-257
mechanical properties 283-289
mechanical properties of blended elemental powder 289
microstructure (F) 282-283
plasma rotating electrode process (PREP) (F) 251-253, 279, 282-283
rotating electrode process (REP) (F) 279, 282-283
screening for oversize particles 253
trial airframe and engine parts and weight comparison (T) 293

Titanium alloys, specific types
Ti-6Al-2Sn-4Zr-2Mo
tensile properties (F) 292-293
use in Pratt & Whitney Aircraft third-stage fan disk (F) 294-295
Ti-6Al-4V
impact properties produced from blended elemental powders (T) 291
microstructure (F) 282-283
prepared by REP (F) 282
tensile properties of blended elemental alloys (F) 290-291
Ti-6Al-4V (Extra Low Interstitial)
tensile and toughness properties (T) 285-287
use in F-16 horizontal stabilizer pivot shaft (F) 293-294
use in F-18 arrestor hook support fitting (F) 293-294
Ti-6Al-4V (Standard)
fatigue properties (F) 284-286
tensile and toughness properties (T) . . . 284
use in Boeing 747 walking beam support fitting (F) 293-294
use in compressor rotor for Williams International F-10 engine (F) 294-295
use in General Electric spool for the TF-34 engine (F) 294-295
use in McDonnell Aircraft drop-out link (F) 293-294
Ti-6Al-6V-2Sn
effect of heat treatment on tensile properties (T) 289
fatigue endurance (F) 286, 288
heat treatment conditions (T) 286-288
impact properties produced from blended elemental powders (T) 291
tensile and toughness properties (T) 286-287
tensile properties of blended elemental alloys (F) 290-291
use in F-14A nacelle frame 294-295
use in fuselage braces by ceramic mold process (F) 292-293

Tolerances 50-51
for ferrous parts of uniform cross section (T) 51

Tool steels 113-119
cutting performance 115
extrusion plus hot work 118
grinding (F) 115-116
high-temperature sintering 118-119
hot isostatic pressing plus hot working . . . 118
processing 118-119
properties 114-115

Tool steels, specific types
M2S
out-of-roundness distortion (F) . . 115, 117
comparison of carbide size distributions with conventional bars (F) 113-114
T-15
comparison of carbides with wrought (F) 114-115

Tooling (F) 45-47

Toughness
of titanium alloys (T) 284, 287

Tungsten-rhenium. *See* Rhenium-tungsten

Turbine disks ((*see also* Disks)
mechanical properties of René 95 (T) . . . 269

Turbine-engine components
use of nickel superalloys (F) 275-277

Turbine wheels (*see also* Dual-property turbine wheels)
mechanical properties of René 95 (T) . . . 268

U

Upsetting
effect on mechanical properties (F) . . 97, 99

V

Vacuum atomization
for nickel superalloys (F) 249, 252
Vacuum sintering
of stainless steels 132-134

W

Water atomization (F,T) 146-149
for preparation of ferrous powders 60-61
for preparation of stainless steels .. 123, 125
Wear resistance
of bronze structural parts 167
Weight
of aluminum alloys (F) 215-216
of titanium alloy airframe and engine parts (T) 293
Welding
conditions for various applications in GlidCop (T) 175